TODAY'S HOMESTEAD
Volume II

Dona Grant

Cover photography by Karl Grant.

"Today's Homestead Volume II," by Dona Grant. ISBN 978-1-60264-273-7.

Published 2008 by Virtualbookworm.com Publishing Inc., P.O. Box 9949, College Station, TX 77842, US. ©2008, Dona Grant. All rights reserved. No part of this publication may be reproduced, stored in a retrieval system, or transmitted in any form or by any means, electronic, mechanical, recording or otherwise, without the prior written permission of Dona Grant.

Manufactured in the United States of America.

TODAY'S HOMESTEAD

HOW TO LIVE AN OLD FASHIONED LIFESTYLE IN THE MODERN WORLD

VOLUME II: IN THE GARDEN

TABLE OF CONTENTS

CHAPTER ONE: In The Garden 1
CHAPTER TWO: Vegetables Galore...... 40
CHAPTER THREE: The Herb Garden. 196
CHAPTER FOUR: The Berry Patch 228
CHAPTER FIVE: In The Orchard 266

CHAPTER ONE
IN THE GARDEN

GARDENING IS AN INTEGRAL part of homesteading. The main purpose of gardening is to enrich your quality of life on your land.

Perhaps your garden is a full acre in size and abounding with every vegetable imaginable. Maybe your garden is a couple of raised beds in the back yard. Do you have tomatoes and peppers growing in pots and planters on your deck, awaiting the day when you can have a real garden? How about an herb garden tucked neatly outside the kitchen door? Are half of all your houseplants really herbs, neatly potted and sweetly perfuming your home? It doesn't matter how or where you garden as long as you do.

The joys to be found in producing your own luscious tomatoes, hefty pumpkins, sweet-as-honey melons, and spicy peppers will have you coming back year after year for more. But it is not only the final product that keeps the homesteader planning out that garden every spring, it is also the pleasure we take in the work itself. Just being outdoors in the sunshine, birds serenading you while you work the soil or plant your seeds, the smell of fresh turned soil, or the aromas of sage and rosemary heavy in the air around you.

The excuses for not gardening are equally many. Too little time, too heavy or too sandy soils, or not being able to do the physical work involved in gardening. Every one of these excuses can be overcome fairly easily. Take each one individually for a moment. Not enough time to garden? You can garden successfully with a very minimal time investment. If you don't have a rototiller, hire a neighbor to come turn your garden soil for you.

It takes only a few minutes to plant store-bought vegetable plants such as tomatoes and peppers, cabbage and broccoli. It very well may take a half an hour of your time to plant an entire patch of green beans, peas, or corn. Most of us spend more time then that in front of the television or the computer.

Is your soil poor? Too heavy? Too sandy? Those are faults easily overcome by the addition of organic matter to your soil. There are also crops that like sandy soils, such as watermelon, and then there are other crops that prefer heavier, new soils, such as potatoes. Within one year, with the right soil additives, you can change unsuitable soil to rich garden loam. Of course, all soils require yearly maintenance to keep them at their best. This maintenance is easy and can be accomplished with very minimal time or cash investment.

If you feel that you are not up to the physical work involved in working the conventional garden, please bear in mind that there are many alternatives that keep even wheelchair bound gardeners happy and working. These may require a little more cash outlay initially, but the return in produce, security, and pleasure will be well worth it.

So, you see, anyone and everyone who wants to grow a garden and be more self sufficient can do so. There are a few things you will need to know as well as some choices you will have to make to get started.

Some homesteaders strive to live an organic lifestyle while others are looking more for the self-sufficient, natural existence. Organic gardening is more labor intensive and to some degree, healthier.

Some methods of organic pest control involve natural, organic pesticides such as insecticidal soaps, oils, and hand picking.

Garden soils

View your garden's soil as the foundation of your food storage plan. The better your soil, the better your garden, the fuller your store room will be. Soil is a complex mixture of many differing ingredients, which contains all the nutrients that sustain life on earth. Plants are especially adapted to extract these nutrients and convert them into useable forms. Our jobs, as gardeners is to keep our soil in the best possible condition by replacing nutrients that the plants use up.

The components of soil are: inorganic matter such as rock and mineral particles; dead and decaying organic matter; water; air; and all manner of living organisms such as insects; bacteria; fungi; earthworms; protozoa; and viruses. The quality of garden soil is determined by the proportions and quality of each individual component in that soil.

Soil varies widely not only from place to place but also within the same space, at different depths of the garden site. Your garden site is made up of layers that differ in composition, therefore differ in texture and color as well.

Top soil is the upper-most layer of your soil. It is darker than the deeper layers because it is richer in organic matter. This is where most of your living organisms reside and where most of your soil nutrients will be found. Depending on conditions, topsoil may be only an inch or two deep, or it may be several feet deep. This topsoil is what good gardeners try so vigilantly to build up. The deeper your topsoil, the healthier your plants will be.

Under the topsoil layer is the subsoil layer. Subsoil is more difficult to dig, and because it usually has a high clay content, will be stickier to work with when it is wet. Often your subsoil will have a distinct color, such as red or yellowish, caused by the mineral particles that are present in this layer. Hardpan is a type of subsoil that is so full of these mineral particles that it hardens to the point where it interferes with drainage. It can also make it impossible for tree and plant roots to penetrate the soil layer.

Beneath the subsoil layer is the geologic base. Depending upon the area you live in, and the make-up of that area, the geologic base can be solid bedrock, or it can be light, sandy, and porous.

Your soil's texture is classified as either sand, silt, or clay. Most soils are a mixture of the three, however, it is the proportions of these three particles that determines what type of soil you actually have.

Sand feels gritty when a bit of it is rubbed between your fingers and thumb. Sand will not hold together very well, even when wet. Sand is the largest soil particle. Large sand particles are called gravel. Sandy soil is easy to work. Many years ago, people called sandy soil, light soil, because it could be worked with a light team of horses. Sandy soil drains easily, but the nutrients that are so important to plant health drain away with the water. Sandy soil takes constant addition of organic matter to build it up and maintain nutrient levels.

Silt particles are smaller than sand but larger than clay particles. Silt feels smooth and floury when a bit of it is rubbed between the fingers and thumb. Silt also packs together slightly better than sand does, therefore, it drains slower and washes fewer nutrients away with the water. As silt dries out, it become light and powdery. Silt requires the addition of generous amounts of organic matter to lighten it up and encourage organic activity.

Clay is the smallest soil particle. When a bit of moist clay soil is rubbed between fingers and thumb it can be molded into a wormlike cylinder. Clay soils pack together well, and when dried will form hard, stony lumps. Clay particles actually absorb moisture and nutrients, swelling up in the process, closing soil pores and compacting your soil. This hampers drainage. The soil will shrink again as the clay dries, but it remains hard packed and will also form deep cracks along the surface. Clay soils need the addition of large amounts of organic matter to lighten the soil up and encourage organic activity.

Loam is the name given to medium textured soil that contains sand, silt, and clay particles in fairly well-balanced proportions. Loam that leans a little heavier on the sandy side is often called a sandy-loam. Clay loams contain more clay particles then silt or sand. Loam soils are friable, which means that any large clods will break down easily into smaller particles. A bit of moist loam rubbed between your fingers and thumb will become a dark smear on your fingers. Loam holds moisture and nutrients well, and a good loam encourages organic activity that makes the majority of your soil's nutrients readily available to your plants. Organic matter and fertilizer must be still be regularly added to the soil to maintain the desired level of acceptability.

Organic material

Organic material is derived from both plants and animals. It breaks down in the soil to a gummy, dark substance called humus, which is not only very rich in nutrients but is also an excellent remedy for badly textured soil. Soils that contain a great amount of humus are called rich soils while leans soils are deficient in humus.

Animal manure, straw, sawdust, dead leaves, pine needles, grass clippings and kitchen scraps are all good sources of humus. It is very important that these wastes be at least partially composted before they are added to the soil. Before composting, animal manures contain weed and grain seeds, as well as being too "hot" (too much ammonia content) to put directly onto the garden soil.

If you plan on adding straw to your garden be sure that it is indeed straw, and not hay. Hay contains millions of weed seeds just waiting for the opportunity to sprout forth. Straw is actually the stems left after oats, wheat, or other grains have been harvested, therefore will contain few if any seeds. It is still desirable to compost straw, as in any organic material, before adding it to your garden soil.

Although composted materials contain a fair amount of nitrogen, raw organic materials heat up as they compost, using nitrogen in the process. If you add organic matter to your garden before fully composting it, it will heat your garden soil up as well as using the immediately available nitrogen. Also, many plants do not act favorably to the higher temperature involved with the addition of partially composted organic material.

Always check your soil's pH before planting in the spring. The addition of new compost may slightly alter the ph of your soil, especially if pine needles, shavings, or sawdust have been added.

The easiest, least expensive way to get good quality compost is to make it yourself. Most homesteads actually have everything needed to create wonderful black compost right there on the property.

Composting is the best way of returning to the soil as humus a great deal of our homesteads' organic materials that would otherwise just be wasted. Creating a compost pile is not difficult, and really only requires an investment of time and energy. The best way to make a compost pile is to layer organic matter between layers of rich topsoil. With sufficient water and air a well made compost pile will reduce organic materials such as leaves, grass clippings, kitchen scraps, sawdust, and even newspapers, to usable, rich humus in a matter of a few short months or even weeks.

You should never add sticks, meat scraps, or anything greasy to your compost pile. Sticks take much too long to decompose to be of any use in the making of humus for the garden, and meats and fats will attract all kinds of unwanted scavengers to your compost pile, as well as contributing some undesirable bacteria.

While you can build a successful open compost pile in a secluded corner of your garden, most people find that enclosed piles work better to deter stray dogs, raccoons, etc. from digging in and spreading the compost around. Free ranging poultry also love to get in and scratch for tidbits in an open pile. You can use any type of fencing material for your enclosed pile, from wire fencing, to wood fencing. Just be sure to allow yourself and your wheelbarrow access to the pile!

To make a compost pile, start with a 1-foot thick layer of grass clippings, leaves, or hay laid in a 5 by 5 foot square pile, or slightly larger if you have the room and the materials. Water this pile well. Add a 1-inch thick layer of manure, and top it off with a 2-inch thick layer of topsoil. Gradually add organic matter as it becomes available on your homestead. Tread down and water each layer when it reaches a foot thick, covering it with another layer of topsoil.

Keep the pile moist, especially in dry areas and in the heat of the summer, watering every two weeks or so. It should take no more than three months during the summer to create good, sweet smelling, nearly black compost, and somewhat longer in the winter.

You may want to turn your pile with a pitchfork after letting it work for about a month and a half. When your pile is of a sufficient size, begin a second pile rather then continuing to add to the original pile. If kitchen wastes and hay are not completely composted they will sprout as weeds and volunteers in your garden. So you want to let one pile completely work to compost without adding anything to it for the last month or so.

Animal manures

Along with the compost pile, don't overlook that pile of well seasoned manure that's been building up outside the barn or chicken house. Manure is a valuable, easily renewed resource on any homestead. Manure is high in nitrogen, potassium, and phosphorus, as well as containing many important bacteria that are essential in converting organic matter into humus.

You can grow beautiful gardens very successfully with no other soil additive other than common homestead manures if you follow a few simple guidelines.

Manure should season for a year or so before being added to your garden. Never put fresh, un-composted manure onto your garden soil. Raw manure, especially poultry manures tends to quickly release nitrogen compounds and ammonia, which can burn plant roots and hinder seed germination. Raw manure is also not good for unplanted seedbeds because of the weed seeds raw manure contains.

Because your manure pile will season for a longer period of time then will your compost pile, you will need to turn your manure pile every month or so to ensure even, complete composting. A hayfork or pitchfork comes in very handy for this job.

Most homestead animal manure can be successfully used on the garden. Sheep, goat, cattle, horse, and rabbit manures are all commonly used. Poultry manure is considered a hot manure, and needs to be composted very well before using. Hog manure also needs to be very well composted before incorporating into your garden site.

Never use the manure of cats, dogs, or

humans in any fruit, vegetable, or flower garden that is to be cultivated by hand. The potential for disease and parasite transmission is too great compared to any possible benefits from using any of these manures.

Green manure

Let's consider green manure for a moment as well. Green manure is simply a crop which is grown for the sole purpose of plowing it under in an effort to incorporate it into the soil to increase soil fertility. This is also a good way to increase the organic matter and humus content of your soil. Green manure uses quick growing crops such as winter rye, rape, vetches and clovers. Green manure should be tilled into the soil before they have a chance to flower and produce seeds.

If clover or vetch is planted with winter rye or rape you will eliminate the need to add nitrogen to your soil as you will have to do when planting either winter rye or rape alone. The use of green manure is becoming more popular with homestead gardeners every year.

A green manure crop that is planted in the fall and allowed to grow all winter before being plowed under in the spring also serves as a cover crop, protecting the soil from erosion by wind, rain, and snow run-off.

Some gardeners make planting green manure a part of their annual crop rotation, planting it selectively in the garden during the actual growing season. As certain vegetables are harvested, simply plant your green manure crop in their place. Plow the entire garden in after all crops are harvested in late fall, when the green manure crop is sufficiently mature.

Green manure crops

Legumes can be grown alone or in combination with other non-legume crops. These are:

- Alfalfa (medic ago sativa) is a perennial plant that is a significant nitrogen contributor. It needs good drainage and a pH that is higher then 6.5.
- Red clover (trefoil pratense) is a biennial that grows quickly and is often incorporated into the soil in the same season it is grown. Contributes nitrogen and phosphorus to the soil. Fairly tolerant of poor drainage and acid soils.
- Dutch white clover (trefoil repens) is a perennial that tolerates droughty soils and adds nitrogen to

the soil. Fairly tolerant of acid soil.

- Sweet white clover (Melilotus alba) is a biennial that produces an extensive root system that accumulates phosphate from rock powders in the soil. Needs alkaline soils and good drainage.
- Hairy vetch (vicia villosa) An annual with good nitrogen capture. Tolerates moderate drainage and makes a good winter cover crop, especially when planted with rye.

Non-legumes should be grown in combination with legumes for best results, but can be grown alone. These are:

- Annual ryegrass (lolium multiform) Is an annual grass that provides fast cover. It tolerates a wide range of soils and growing conditions.
- Oats (avena sativa) is an annual that provides quick cover and prefers well-drained, loamy soils. Will tolerate some acidity, but prefers more alkaline soils.
- Rape (brassica napus) An annual that prefers moderately well-drained loamy soil.
- Winter rye (secale cereale) A very winter-hardy annual that prefers

well-drained soils.

Many homestead gardeners still prefer to allow their soil to rest for the winter before planting again in the spring rather the planting green manure. They rely instead on the liberal addition of compost and manure. Use whatever method works for your soil type and your lifestyle.

A healthy garden
You can minimize insect pests and disease organisms in your garden by following a few easy steps. It all starts with soil that is rich in humus, drains well, and has ample nutrients. As stated earlier, amend your soil regularly with compost and apply organic mulches continually to enrich the soil and retain moisture.

Test soil samples for nutrient levels and pH, and make any necessary adjustments. Apply only as much fertilizer as you need, since excess can cause growth imbalances that weaken plants and invite insects and diseases.

Cull any plants that are diseased to prevent the pathogens from spreading throughout the garden. Look for vegetable varieties that are resistant to wilts and other diseases. Disease resistance is noted in seed catalogs, on pot labels, and on seed packets.

Keep a close eye out for insect pests, especially aphids and leafhoppers, which can spread diseases as they feed. Their feeding can also create entry points for rot organisms. Plants that are weakened by insect pests are more susceptible to disease organisms.

If you allow weeds to get out of hand in your garden, the competition for nutrients and moisture can also weaken your crop plants, leaving them susceptible to infection with disease organisms. Crowding by weeds or close planting also reduces air circulation, allowing moisture to remain on plants long enough to allow fungal and bacterial pathogens to infect plants.

Under-watered or over-watered plants are both vulnerable to disease. Disease organisms spread more rapidly among wet plants, so drip irrigation is the most efficient way to water. It doesn't moisten plant leaves and stems, it saves water, and it puts water right in the root zone where plants need it most. If you do use overhead sprinklers, water during the early morning so your plants have time to dry before nightfall. Also, remember not to cultivate or harvest your garden crops when the plants are wet to avoid spreading disease organisms.

Since some diseases are soil-borne, planting related crops in different areas each year improves their chances of avoiding insect infestations and diseases and reduces the amount of the organisms living in the soil. At least a three-year rotation with any plant relatives is recommended.

At the end of the growing season, if plants are healthy, spade or till plant residue into the soil, or pull it and add it to your compost heap. This simple clean-up task helps to remove places where pests can over winter in your garden, and adds organic matter to improve the soil. If plants suffered from heavy insect infestation or disease, it's better to burn the plant residue, feed it to chickens, hogs or other livestock, or bury it in a spot far from the garden.

Lasagna gardening

Lasagna gardening is a form of layered gardening in a standard type of garden.

There are no hard and fast rules about what to use for your lasagna garden's layers, just so long as it's organic and, of course, doesn't contain any protein (fat, meat, or bone). The benefits to lasagna gardening are many.

- You don't have to remove the sod or do any extra work, like tilling the soil, or removing weeds or rocks.

- Your soil stays rich in nutrients and moisture retentive, yet maintains good drainage.
- You get the benefits of a raised bed garden combined with the benefits of a conventional garden site.

Creating a lasagna garden is easy once all the materials are brought together in one place.

Measure out and mark the area for your garden using a string or long rope to get the desired shape. Then, cover the area you've marked with wet newspapers, overlapping the edges using five or more sheets of newspaper per layer. Cover the newspaper with one to two inches of peat moss or other partially composted organic material such as dead leaves, straw, etc. Next, layer several inches of partially composted organic material, such as grass clippings, kitchen scraps, manure, etc on top of the peat moss.

Continue alternating layers of peat moss and organic material until desired thickness is reached. Water this lasagna bed until it is the consistency of a damp sponge. Plant as desired, and mulch well to help retain moisture.

You need less loose material to plant in than you might think. When you lay the first few whole sections of wet newspaper, lay it directly on top of the ground, sod, tree roots, and all. Just cover whatever is there. When you layer your garden, layer additions of one to two inches of grass clippings, two inches of peat moss, one to two inches of compost, and a topping of peat will give you a total of about six to eight inches to plant in. After you harvest, you'll see that earthworms have moved in and mixed things up a little.

Each fall, add more compost, grass clippings, shredded leaves, etc. to the lasagna plot.

To prepare the garden for another year of planting in the spring, spread additional compost onto the plot, and add more clippings, peat, manure and other compost You'll see your garden gain a few inches in height each year.

Soil pH

Soil acidity or alkalinity are measured on the pH scale. This scale runs from 0, which is pure acidity, to 14, which is pure alkali. The neutral point is in the middle, around 7. Slightly acid soil, or about pH 6.5 is ideal for most plants. There are a few exceptions, such as blueberries which requires soil in the pH 5.0 range, and cabbage, which does best in soils that are slightly alkali, about pH 7.5. Very few plants will survive in soil more acid then pH 4.0, or more alkaline then pH 8.

It's a fairly easy thing to adjust your soils pH. If your soil is too acid you can add finely ground limestone to your garden to increase alkalinity. Limestone, or agricultural lime, is available at farm stores and garden supply centers. Simply follow the basic directions on the bag for use. Some people prefer to lime their soil in the fall, others like to do it early in the spring. Because the addition of organic material will naturally increase the acidity of your soil to some extent, its a good idea to test your soil every year.

For soil that is too alkaline you can purchase iron in tiny pellet form. Again, follow the instructions on the package for its use. You can also use flowers of sulfur which, over the course of a few months will help add acid to your soil. Both of these products are available at agricultural stores and garden supply centers.

Be careful not to overdo your soil pH adjustments. Too much lime, or too much sulfur will over adjust and you will have the opposite problem from the one you started with. Use only as much of each corrective additive as is needed to bring your soil into the desired range, based on soil testing. Soil test kits are also available at garden supply centers, just be sure you spend the money to get a good reliable kit.

Gardening tools

It doesn't take a large financial outlay to get started gardening, however, there are a few tools that will make the job much easier, and less time consuming. Always purchase the best tools your money can buy. Cheap tools will cost more money in the long run. After you've purchased your tools, remember to take care of them. Don't leave them out in the weather, and clean them off after each use. This will extend the life of your tools, giving added value for the price you've paid for them.

At the end of each growing season, its a good idea to go over your tiller and tighten up any thing that may have come loose, remove rust spots, and apply a drop of oil to any moving parts. At the same time, rub down all metal parts of other garden tools with an oily rag before storing them away.

If you have a large area you intend to garden, you'll probably want to invest in a rototiller. A good rear-tine tiller is the best investment, but any tiller will make deep cultivation fast and nearly effortless.

A small tiller can be purchased as well, and is wonderful for cultivating between rows. It will make weeding your garden a snap.

A garden rake and hoe will also be needed, as will a hand cultivating set, and a good length or two of garden hose equipped with a sprinkler head or an adjustable nozzle.

If you only have a small area to cultivate, you may find that turning the soil by hand is a better choice financially. Be aware though, that hand turning soil can be tedious, back tiring work. To turn your soil by hand you will want to invest in a good sharp-pointed spade, and a garden rake to break up smaller clods and rake the soil smooth. If your soil is particularly hard, you may want to invest in a pick or mattock to break up large, hard packed clods. You will also want to purchase a garden hoe, as well as a hand cultivating set, and a length or two of garden hose equipped with a sprinkler head or an adjustable nozzle.

The garden site

Any garden site will have to meet a few basic requirements. The site should be an open area with good drainage, plenty of sunlight, and a good supply of water nearby.

Most vegetable plants require at least six hours of sunshine daily. Only root crops and leafy vegetables can be grown successfully in less then six hours of daily sunshine.

Trees and shrubs should be cleared from within 20 feet of your garden because they tend to be heavy feeders and water users. Any tree that casts shade onto your garden site should be cleared out.

Remember, the size of your garden is not as important as is the quality of your garden soil, the amount of sun received daily, and the availability of water. A small, well-planned garden of only 300 square feet can provide a full year's supply of vegetables for a family of four.

Planning the garden

A measuring tape, notebook and pencil are the only tools you'll need to plan out your garden. Measure your garden site's dimensions and write them in the notebook. Then, draw out a smaller scale model of your garden using a half inch or so to the foot scale.

Be sure to leave sufficient space for walkways. If you plan on using your tiller to cultivate between rows, be sure to leave sufficient space between rows to accommodate your tiller's width. This is where a smaller eight inch width mini-tiller comes in handy. You can maximize your garden space while still enjoying the benefits of mechanical cultivation.

Make allowances for those crops that mature early, such as peas, spinach and lettuce, and plan on putting in a second crop after the first one is harvested. This is called succession planting. You don't want to let any of your garden space go to waste. You can also co-plant to take advantage of all available space.

Co-planting is the practice of using companion plants to fill in the empty space after the initial crop has been harvested. An example of co-planting is English peas and winter squash. Plan on putting several hills of winter type squash between your rows of peas. The peas are planted earlier, and will mature faster then the squash. After the peas have stopped producing and you have pulled the vines, the winter squash vines will be free to run throughout that space, thereby making double use of the area. This is the best way to make double use of the same space where growing seasons may be short and therefore won't allow for a second crop.

If you are short on garden space there is a method of planting known as the three sisters that may help to solve some of your garden space dilemma. In this planting method pole beans and winter squash are planted around corn, which acts as a natural trellis on which the others grow and climb.

Some crops do better in hills then in rows. Squashes, pumpkins, cucumbers, cantaloupe and watermelon are all often planted in hills. Hills give you a deeper layer of topsoil for these heavy feeders, and deeper topsoil holds water longer.

When you plan out your garden on paper, keep in mind the direction the sun will be in at various times of the day. You don't want to plant several rows of corn, only to have it come up beautifully, casting a heavy shade on all your tomatoes and peppers. Plant your corn rows near one edge of the garden where any shadows will be cast outside of the garden. Keep this in mind for climbing plants as well, such as cucumbers, pole beans, etc. Trellises can cast a great deal of shade across your garden when they are thick with greenery.

There are also some plants that thrive on less sunlight then most garden vegetables. Spinach and lettuce will both do well in a partly shaded area. So if you happen to have an area of your garden that doesn't get daylong sunshine don't despair. The space will not be wasted if you plan on putting your leaf crops in that area.

Keep in mind the average yield of each

vegetable you want to plant. Some vegetables will give an enormous yield. Radishes and lettuce are two vegetables that are usually eaten fresh only, and one long row of each will give you more salad greens and radishes than your family could possibly eat. Much of your crop would be wasted. It is much smarter to plant shorter rows of each, making successive plantings every three weeks if they are desired. Green beans will often produce huge yields, but green beans can be frozen or canned for future use, so large plots of green beans can be used easily.

Be sure to plan on planting some natural insect repellents along with your vegetables too. A border of marigolds or even just planting a few of them in amongst your vegetables will help to deter a multitude of insects from invading your garden. Chrysanthemums are another effective plant to use for natural pest control in the garden.

One of the most important things to keep in mind as you plan your garden is your frost dates. Last expected frost date is the average date in the spring that you can expect a frost in your area. This date is important because it serves as a guide for spring planting. Likewise, the earliest expected frost date is the average date in the fall when you can expect your first frost for your area. This date is also important to keep in mind because most vegetables that have not been harvested prior to the first autumn frost will be damaged or ruined entirely. Using these two dates together can help in planning what types and varieties of vegetables to grow depending on the length of your individual growing season. It is very difficult to try to grow a crop that has 140 days to maturity in a 120 day growing season. So be sure your vegetable selections are appropriate for your growing area and season.

These planning instructions are the same if you are laying out a large vegetable garden, or a smaller herb garden. Perhaps you will want to combine the two and have a corner of your vegetable garden dedicated to growing your favorite herbs.

One word of caution, however. Many perennial herbs have a tendency to spread freely. The mints, in particular, lemon balm, horseradish, and rhubarb will all spread like wildfire if not checked somehow. For this reason, many people like to plant herbs in a separate garden rather than setting aside a corner of the vegetable garden. If you have enough extra space in your garden and you don't care if your herbs ramble, then having them planted along side your vegetables is perfectly fine.

Container gardening

Container gardening is just what it sounds like. Growing garden plants in pots and planters allows many people to garden who might not otherwise have the opportunity due to health issues, soil quality issues, or space availability. Gardening in containers is ideal for the elderly, those with little or no garden space, and those with extremely sandy soils. Many garden plants can also be used indoors as lovely, useful houseplants.

Pots and planters come in all sizes and shapes. Planter colors can be bought to match your home's décor, whether for use indoors or outdoors. There are pots made of resin, wood, plastic, concrete, and clay.

Whatever pot or planter you choose must have sufficient drainage holes. Without adequate drainage, the soil will become waterlogged and the plants will most likely die. Most containers sold today do have sufficient drainage holes in them already. If you have a pot that does not have adequate drainage holes you can drill them yourself. Self-watering containers are also available and are very convenient as well. Contrary to popular belief, it is not necessary to cover the drainage holes with gravel or pot shards. It is far better to use a layer of newspaper set in the bottom of the pot or planter. This newspaper layer will help absorb moisture, making it more readily available to the plant later, while still keeping the roots from sitting in water.

Also, you should keep in mind the size requirements of the plants you want to put in the pot or planter. Purchase a pot or planter that is of sufficient size to accommodate the plant or plants they are intended for. You don't want your plants to become root-bound. Root-bound plants dry out easily, and they simply don't thrive well. Larger containers hold moisture longer, and allow plants roots more room to grow.

When choosing the soil for your pots, remember that ordinary garden soil is often too dense and heavy for successful container gardening. You can purchase a commercial planting mix, or you can create your own mix. Mix equal parts of compost or sphagnum peat, composted pine bark, and perlite or vermiculite. For each cubic foot of this mix add 4 oz. Dolomitic limestone, 1lb. Rock phosphate or colloidal phosphate, 4 oz. Greensand, 1lb. Granite dust, and 2 oz. Blood meal.

When you put plants in your containers, plant them the same way as you would plant them in a garden. Almost any vegetable or herb can be grown in a container. Even fruit trees and shrubs can be container grown in the right conditions.

Maintain your container garden by sufficient watering and fertilizing when needed. Water thoroughly when soil begins to dry out. Do not allow the soil in your containers to dry out completely between waterings. Completely dry soil is quite difficult to re-moisten. Container plants also need more frequent feeding then garden plants. The easiest way to fertilize is to use a fish emulsion, seaweed extract or manure tea. Begin by feeding your container plants once every two weeks, adjusting the frequency depending on the plant's response.

Raised bed gardening

Raised bed gardening is just what it sounds like. It involves growing plants in beds that are raised higher than ground level. Raised beds are used to increase a garden's accessibility, especially for those with health limitations that would prevent them from having a conventional garden.

Raised bed gardening is also a good way to solve problems of poor soil quality, and space restrictions. The beds are usually between 3 and 5 feet across to permit easy access while working in the beds. Raised beds can be any length desired. All garden vegetables, herbs, flowers, and most berry bushes can be grown in raised beds.

Raised beds can be constructed from many different materials. Stone, wood, brick, and concrete block are just a few of the more common raised bed framing materials.

If you are using wood to frame your beds you will want to be sure not to use chemically treated lumber as many chemicals can leach out of the wood and enter your soil. Naturally rot-resistant woods are available. Lumber such as cedar, cypress, or locust is most often used in raised bed fruit and vegetable gardens. You can use less rot-resistant lumber in the construction of your beds, but you will probably find that you have to replace it sooner, or more often then if you used more rot-resistant varieties.

If using brick, concrete block, or stone, be aware that slugs and snails are more attracted to these materials because of the moisture, and cooler temperatures they hold. You may have to be slightly more vigilant at protecting your fruits and vegetables from these pests if using any of these materials to frame your beds.

Soil for your beds can be purchased by the bag from a garden supply store, however, that tends to be a needless expense. It is thriftier by far to buy topsoil by the truckload, and then mix in your own natural amenities such as organic matter and mineral supplements like bone meal, limestone, etc. depending upon your soil testing results.

Proper shaping of raised bed soil within the bed is important. The soil should be mounded in the center, tapering or sloping at the sides. This shape will allow proper drainage and will actually help to conserve water. Any soil that is washed down from the top of the bed will be caught and held in place by the bed frames. In this way, raised beds help greatly diminish soil loss from erosion.

Beneficial insects

Some of our beneficial insects are obvious to everyone. Honeybees, bumble bees, and lady bugs are perhaps the most well known of the beneficial insects. The list also includes various flies, wasps, and beetles that no homesteader would want to be without. Beneficial insects can be attracted to your garden from the wild, or they can be purchased through many garden catalogs and released onto your homestead.

The best way to protect beneficial insects on your homestead is to limit the use of pesticide products to only extreme situations. Even organic, or botanical, pesticides can kill beneficial insect species, so be judicious in their use as well.

You can encourage beneficial insects to take up residence on your homestead by providing the few necessities they have. Keep a shallow bowl with stones and water in it so that small insects can access the water without drowning. Perhaps consider laying stones, or brick about in your garden to attract beneficial insects to the shelter they provide. Allow a patch of weeds to grow outside your garden so beneficial insects have a haven away from the open nature of garden crops. Plant specimens that are attractive to beneficial insects along your garden border, such as catnip, dill, and yarrow.

Don't forget toads and snakes as well in the battle against insect pests. Toads consume an enormous number of insects each season, as do snakes. Snakes also deter mice and other rodents from sampling your garden produce before you do. Most snakes that will be attracted to your garden are not poisonous, however, it is always wise to approach any snake with caution until you are certain what species you have. The same items that attract and keep beneficial insects in your garden will attract toads and snakes.

- Parasitic wasps range from the tiny Trichogramma wasps to the huge black ichneumon wasps. Parasitic wasps lay their eggs inside or on the body of their host insect. Some larvae grow by absorbing nourishment through their skins, while others literally consume their host insect. Parasitic wasps are used in the control of aphids, whiteflies, and some caterpillars.
- Yellow jackets are detested by most people, but not by homestead gardeners. Yellow jackets prey on many caterpillars, flies and other insect pests and take them back to their nests to feed their own developing brood.
- . The ladybug is perhaps the best known of the beneficial beetles. There are actually more then 3000 different species of lady bugs. Not all species of ladybugs are beneficial though. Mexican bean beetles belong to the lady bug family as well. Convergent lady beetles (*Hipodamia convergens*) are the most well-known of the lady bugs, being sold by garden supply catalogs to gardeners across the country.

These insects feed on small, soft bodied pests such as aphids, mealy bugs and spider mites. Both larvae and adult eat insect pests. The drawbacks to purchasing these insects is that they don't necessarily stay on your property, but will disperse great distances unless housed in a greenhouse, and they form large clusters of ladybugs in homes and garages as they prepare to over-winter.

- Ground beetles are medium to large, fast moving, blue-black beetles that hide under stones, boards, or debris throughout the day. At night they prey on cabbage root maggots, cutworms, snail and slug eggs, tent caterpillars, and many other pests. Placing a couple of larger stones in your garden will encourage the presence of these beneficial insects.
- Rove beetles are small to medium insects that resemble earwigs without the pincers. Many species of rove beetles feed on garden pests such as root maggots and cut worms. Other rove beetles aid in the decomposition of manure and plant materials.
- Fireflies are beetles, and the

same delightful beetle that sparks up the night with their cool greenish light also aids in insect control in your garden by consuming other insect larvae. Both the adult, and the larvae of fireflies are beneficial to the homestead gardener.

- Tachinid flies are large, bristly, nearly black flies that lay their eggs on the bodies of cutworms, caterpillars, corn borers, stinkbugs and other pests. Tachinid flies are important aggressors in the fight against tent caterpillars and armyworms.
- Syrphid flies are black and yellow or black and white striped flies that are often mistaken for bees or wasps. They lay their eggs in aphid colonies. The larvae go on to feed on the aphids. The larvae look very much like small, gray, translucent slug-like maggots. Do not mistake them for small slugs. They can be very beneficial to the homestead gardener.
- Aphid midge larvae are tiny orange maggots that are voracious aphid eaters. They can be purchased through garden supply catalogs and insectaries.
- Lacewings are delicate ½ inch to

1 inch size, green or brown insects with large, transparent wings. They lay pale green oval eggs, each at the tip of a long fine stalk in the middle of garden plants. The larvae of lacewings feed on aphids, scale insects, small caterpillars and thrips.
- Spiders are not true insects, but are being included in this section. Some spiders spin webs and funnels to catch insect pests while others lay in wait to pounce on the unwary insect passer-by. Spiders are some of our best allies in the fight against insect pests in the garden and all around the homestead.

USDA hardiness zones for North America

Different plants vary in the temperature variations that they can endure. The USDA has developed plant hardiness zone maps to aid in deciding which plants will grow successfully in the area where you live. The zones are based on the average coldest temperatures for each area.

These maps do not take into consideration other aspects of plant growth such as soil type, rainfall, daytime high temperatures, day length, wind, microclimate, etc. These will also have to be considered in making selections for your gardens, berry patches, and orchards.

These maps are available in seed catalogues, and online, and are meant to be a general guideline only.

CHAPTER TWO
Vegetables Galore

Whether you plant only heirloom vegetables and save your own seed, or like to experiment with all the newest hybrid cultivars, there are hundreds of different varieties of vegetables just waiting to be discovered, or rediscovered. Armed with a little knowledge, and the right tools, any homesteader can provide ample vegetables fresh from the garden to feed an entire family, whatever the size may be.

Planting and harvesting guidelines for vegetables

Asparagus:
   Asparagus is a fernlike perennial vegetable that is grown for it's tender young shoots. Asparagus is rich in B vitamins and dietary fiber.
   Select your asparagus bed carefully.

Asparagus will occupy the same bed for twenty years or more. Asparagus can tolerate some shade, but full sun produces the healthiest plants and helps prevent disease. Asparagus also prefers rich, light, fertile soils.

Fifty to one hundred plants are sufficient for a family of four. You can purchase year old roots, called crowns, or you can plant your asparagus from seed. Both have their advantages and drawbacks.

Planting crowns will give you a year head start over seed grown plants, however, crowns tend to suffer from transplanting shock.. If you plan to plant crowns be sure and purchase them from a reputable nursery, and buy only disease free, one year old crowns. Two year old crowns won't produce any faster, and they have a tendency to suffer transplanting shock even more readily then the one year old crowns do.

Plant your crowns immediately, or wrap them in dampened sphagnum moss until you can get them in the ground. Try to purchase only male crowns if possible. Male plants tend to produce higher yields then female plants because the female plants put a portion of their energy into producing seeds which randomly sprout into new plants. Unless you are looking to sell these plants, you will probably find yourself weeding them out of your asparagus bed anyway. It is better in that instance to plant only male plants and enjoy the bountiful crop they will produce for years to come.

Seed grown asparagus plants take more patience to grow, but they don't suffer from the transplanting shock so common with asparagus crowns. You can buy an entire packet of asparagus seed for the cost of one nursery-grown crown. Also, most seed-grown asparagus plants will go on to out-produce plants started from crowns.

Planting guidelines:

Whether planting crowns or seeds, be sure to purchase asparagus disease resistant cultivars.

In the north, start your asparagus seedlings indoors around late February. It is best to plant single seeds into individual peat pots. Try to maintain a temperature of around 77 degrees, using bottom heat if necessary. When the seeds have sprouted, lower the temperature to about 65 degrees. Once all danger of frost is past, and your seedlings are about 1-inch tall, plant the seedlings outdoors into a nursery bed.

If you want to plant only male plants in your garden, when the tiny flowers appear, look at them closely with a magnifying glass. Female flowers have well-developed, three-lobed pistils, while male blossoms are larger and longer then the female flowers. Weed out all the female flowers.

The following spring, transplant to their permanent bed. Plant the seedlings in trenches approximately 1 to 1 ½ feet apart. Top with 2 inches of good soil and clip off any little branches that are below soil level. In two weeks, add another 2 inches of good soil, and add soil every two weeks until the soil in the trenches is slightly mounded above the surface level to allow for settling of the soil.

In the south you can plant seeds directly into a nursery bed as soon as you can work the ground. Sow two seeds to the inch in rows 11/2-inches apart. Asparagus seeds take about 30 days to germinate, so be sure the nursery bed doesn't dry out during this time. When the asparagus shoots are 3 inches tall, thin them to 4 inches apart. In late summer, around the end of august, transplant the male plants to their permanent home.

To plant asparagus crowns you will need to dig trenches that are 12 inch wide and approximately 7 inches deep, keeping your trenches 4 feet apart. Set the asparagus crowns with their long thin roots spread out and draped over the soil in the bottom of the trench. Keep the crowns about 1 to 1 ½ feet apart. Cover crowns with about 2 inches of good soil. In two weeks, add another 2 inches of good soil. Continue adding soil every two weeks until the soil in the trenches is slightly mounded above surface level to allow for soil settling.

Maintenance:

Apply mulch to the asparagus bed throughout the year to smother weeds. A weedy asparagus patch will reduce your asparagus crop yields. Water your bed regularly during the first two years after planting. After the first two years, asparagus roots tend to go deep into the soil so regular watering is less critical. Fertilize your asparagus in the spring and fall.

Don't clean up your asparagus bed in the fall. Leave all mulch, straw, and dead leaves on the bed over the winter to give some protection from the cold. Remove and destroy any old foliage and mulch before the new shoots appear in the spring. Old leaves and mulch can harbor insect pests and their eggs as well as diseases.

Every spring you should add a mound of about 6 inches of new soil over your asparagus to be sure your crowns are covered. Crowns that work their way to the surface will result in lower yields and less tender spears.

If you want to try growing white asparagus, simply continue to mound soil up as the emerging shoots grow, thereby keeping the spears blanched.

Insects and diseases:

Asparagus pests include asparagus beetles which feed on young spears in the spring and attack foliage in the summer.

Asparagus beetles are about ¼ inch long, and are a metallic blue-black with three white or yellow spots on their backs. They lay dark eggs along the leaves which hatch into light gray or brown larvae with black heads and feet.

Control by handpicking, or spray or dust seriously infested plants with rotenone or carbaryl.

These pesticides are also effective against the 12 spotted asparagus beetle, which is reddish brown with six black spots on each wing cover.

Zigzag tunnels on stalks is a sure sign of the asparagus miner, another foliage-feeding insect. Destroy any infested ferns to control this pest.

Asparagus crown rot is a disease that causes spears to turn brown near the soil line. Prevent crown rot by planting in raised beds, or be sure and maintain good drainage and keep soil pH above 6.0.

Fusarium wilt, a fungus which causes spears, ferns and stems to be small with large lesions at or below soil level, can be controlled with good garden sanitation.

Research has shown that asparagus beds over one year old that have had rock salt applied to the bed resist crown and root rot diseases such as fusarium and crown rot, and improves overall plant health. Be sure to use only sodium chloride (NaCl). Don't attempt to use iodized table salt or calcium chloride (CaCl). Pickling salt is fine to use.

Apply salt at the rate of 2.5 lbs per 100 feet of row just before shoots appear in the spring, but no later than July 4th.

Young asparagus spears can be damaged by late frosts. Young spears turn brown and become soft or withered when touched by frost, so protect your emerging spears by covering the bed with sheets, newspaper, or clean mulch when frosts are predicted.

Harvesting:

Do not harvest asparagus during the first two years after plants are in their permanent bed. During the third season, pick the spears over a four-week period, and by the fourth year, extend your harvest to eight weeks. In early spring, harvest spears every third day or so. As the weather warms, you might have to pick twice a day to keep up with production.

Beans:

All beans belong to the legume family. There are hundreds of different cultivars available to today's gardener. All of these cultivars fall into four general bean types. Bush types are generally self-supporting while pole types require staking, trellising or wiring for support to their twining vines. Runner beans are similar to pole beans, although they do require cooler growing conditions then do pole beans. Half runners fall somewhere in between pole beans and bush beans.

Every homesteader should grow beans for pods, called snap beans, such as wax or green beans, as well as beans for drying, sometimes called horticultural beans, such as Great Northern Beans or Kidney beans.

Most snap beans mature in as little as 40 to 60 days, while dry beans will require 60 to 100 days to reach maturity.

Planting guidelines

Although harvested and used differently, all beans are grown in basically the same manner. Generally speaking, beans are a warm weather crop. The exception to this is the fava bean, which requires a long, cool growing season. Most beans grow best at air temperatures of 70 to 80 degrees F, and soil temperatures of at least 60 degrees F. Wet, cool soil will only cause bean seeds to rot.

Beans need a sunny, well-drained area that is rich in organic matter. Lighten heavy soils with extra compost to aid the seedlings in emerging.

Plant approximately 12 bush bean plants, or 4 hills of pole beans per person.

Beans are self-pollinating, so you can grow cultivars side by side with little danger of cross-pollination. If you plan to save seed from your plants, however, separate cultivars by at least 50 feet.

Bean seeds can remain viable for three years, and show a germination rate of about 70 percent. Don't soak or pre-sprout bean seeds before sowing in the garden.

Bean seeds can be dusted with an anti-bacterial inoculate powder called Rhizobium phaseoli. This inoculate should be available at your local garden center or greenhouse. It Is not a vital step, but it does offer some protection against some bacterial disease organisms.

Beans generally don't transplant well, therefore should be sown directly into the garden soil. Only in the shortest of growing seasons should indoor seeding for later transplanting be attempted, and then all diligence and care should be given during transplanting.

Outdoors, plant the first crop of beans two weeks after the last expected frost date for your area. Sow seed one inch deep in heavy soils and 1 ½ inches deep in lighter soils. Firm the earth over the seeds to ensure soil contact.

Space most bush cultivars 3 to 6 inches apart in rows spaced 2 to 2 ½ feet apart. Bush beans will produce the greatest majority of their crop over a two to three week period. For more continuous harvest, stagger successive plantings at two week intervals until about two months from the first expected fall frost date.

Bush beans germinate in about 7 days. Keep soil moist throughout the germination period, as well as the period just before plants blossom.

Pole beans are even more sensitive to cold then are the bush beans. Pole beans also take longer to mature then bush beans, however, they produce about three times the yield of bush beans in the same garden space, and they bear continually until the first killing frost in the fall. Plant pole beans three weeks after the last expected frost date.

Space pole beans 10 inches apart in rows that are spaced 3 feet apart. Seeds should be sown 2 inches deep in compost rich soil. Pole beans can also be planted in hills around a tee-pee-type of pole frame. Space hills 3 to 5 feet apart, sowing 6 to 8 seeds per hill. Thin to 4 strongest plants per hill when seedlings emerge. Provide supports at planting time or as soon as the first two leaves of seedlings emerge. Pole beans germinate in about 14 days. Keep soil moist throughout germination period as well as the period just before plants blossom.

Maintenance

Beans do not need added nitrogen during the growing season. Being legumes, beans produce nitrogen via bacteria that thrive in nodules on bean roots. Adding extra nitrogen to your beans will only result in lush green foliage growth with little to no blossom or bean production.

A little extra potassium does help increase bean growth, especially with the heavy feeding lima beans.

Do not cultivate or pick beans during wet or overly humid weather when the bean leaves are wet. Working in wet or damp beans is a sure way to spread bean disease spores.

Keep your beans picked over the length of the growing season. Beans produce the majority of their crop over a 2 to 3 week period, but will continue to produce beans at a reduced rate for several weeks longer.

After your beans have stopped producing, pull out plants or vines, and any temporary supports. It is a better idea to use temporary supports rather then permanent ones so that you can rotate your bean crop every year, cutting down on the possibility of disease.

Insects and diseases

Insects that attack beans include aphids, Japanese beetles, Mexican bean beetle, as well as corn ear worm, cabbage loopers and European corn borers.

Aphids are tiny insects having long antennae and two tubes protruding from the abdomen. They are usually green in color, thereby blending in well on their host plant.

Aphids suck plant juices causing foliage to shrivel and be distorted which eventually leads to foliage drop. Aphids excrete honeydew which encourages sooty mold growth and encourages the presence of ants, which feed on the honeydew. Control by using insecticidal soap, or products containing carbaryl, pyrethrim or rotenone. Releasing lady bugs, lacewings, or aphid midges will also help with continued control of aphids.

Japanese beetles are metallic bluish-green beetles with bronze wing covers. They are about ½ inch long. Japanese beetle larvae are fat, white grubs with brown heads. Adult beetles chew leaves, leaving behind mere leaf skeletons. Japanese beetles will chew blossoms and can completely defoliate a plant. Larvae feed on plant roots. Japanese beetles can be controlled by hand-picking of adults, by applying floating row covers to plants, by application of milky disease spores to garden soil, or by installing baited traps throughout the area. Rotenone is also effective on Japanese beetles.

Mexican bean beetles are yellowish-brown beetles, about ¼ inch in size, with 16 black spots on wing covers. Larvae are flat, dark yellow grubs with long branched spines. Adults and larvae chew on plants from the underside of leaves, leaving a characteristic lacy pattern. Plants become defoliated and die. Apply floating row covers to control, or plant beans earlier in season. Hand pick beetles, or release spined soldier bugs. Spraying with neem, pyrethrins, carbaryl, or rotenone also helps in controlling this pest.

Striped cucumber beetles are ¼ inch long, yellowish-orange bugs with black heads and three black stripes going down their backs. These pests can spread bacterial blight and cucumber mosaic. Apply a thick layer of mulch beneath the plants to discourage them from laying their orange eggs in the soil near the plants. Control adults by handpicking, or release soldier beetles, machined flies, and beaconed wasps. Interplanting your beans with catnip, tansy, radishes or nasturtiums repels these pests. Carbaryl dusting helps keep cucumber beetle numbers down as well.

To minimize disease problems, buy disease resistant cultivars, and disease free seeds. Rotate your bean crop every year or two. Do not harvest or cultivate beans while plants are wet.

Common bean diseases include Anthracnose, a disease that causes black, oval, sunken cankers on pods, and stems, and black spots on leaf veins.

Bacterial blight causes large, brown blotches to develop on leaves. Foliage may fall off and the plant then dies.

Mosaic virus causes yellowed leaves and stunted growth. Control aphids and cucumber beetles, which spread the virus.

Bean rust causes reddish-brown spots to appear on leaves, stems, and pods.

Downy mildew causes fuzzy white patches on pods. Lima beans are especially susceptible to down mildew.

If any of these diseases strikes your bean crop, pull and destroy all infected plants immediately. Don't touch other plants with unwashed hands or garden tools. Don't sow beans in that area again for 3 to 5 years. Good plant hygiene, insect control, and proper cultivation is the best preventative against disease.

Harvesting beans

Pick green or wax beans when the pods are pencil-size, and tender. You don't want to leave the pods to form hard bumps on them. Harvest your beans daily to encourage production. Plants will stop producing and die if beans are left on the plants to fully ripen.

Don't pull the pods off the plants. You may pull off an entire flowering head or even uproot the entire plant. Instead, pinch off bush beans with your thumb nail and fingers. Use scissors to harvest pole and runner beans.

Serve or process green or wax beans the same day you harvest them for best flavor. They will hold for up to a week in the refrigerator.

Harvest shell beans for fresh eating when the pods are plump but still tender. The more you pick, the more the plants will produce. Shell beans can be held for up to a week in the refrigerator.

To dry beans, leave the pods on the plants until they are brown and the seeds rattle inside the pods. Seeds should be so hard that you can barely dent them with your teeth. If the pods have yellowed, but not dried completely, and a rainy spell is forecast, cut the plants off near the ground and hang them upside down in a warm, dry place to finish drying. Put shelled beans in airtight containers, and store in a cool, dry place. Dry beans will keep for a year or more.

Beets

Beets are a very versatile and nutritious crop, and one the sensible homesteader will not overlook in garden planning. Beet plants yield delectable greens that contain vitamins A and C, and more iron and minerals then spinach. Beet roots can be eaten fresh, canned and even pickled. Beets are rich in potassium and also contain protein, fiber, iron, calcium, phosphorus, niacin, and vitamins A and C.

Beets produce best in full sun, however, they will tolerate partial shade. Too much shade will produce spindly plants with little or no root globe growth. Beets like a loose, rich, rock-free, soil. Avoid the use of fresh manure where beets will grow to avoid forked roots. Like most root crops, beets do best in well-tilled soils, and will benefit from having the soil cultivated into hills around the growing plant roots.

Planting guidelines

Beets are a cooler weather crop and prefer temperatures in the 60v-65 degree F range.

Sow the unusual seed directly in the garden about 4 weeks before the last expected spring frost date in your area, or as soon as your garden soil can be worked

Sow beet seed ½ inch deep, spacing seeds 2 inches apart. Plant in rows that are spaced 1 foot apart. Thin beet seedlings to 4 to 6 inches apart. The tiny plants removed can be transplanted to another spot, or you can enjoy the delicious young greens in salads, etc. For a continuous beet crop, plant successive plantings every two weeks until the weather begins to turn hot.

Maintenance

Diligence in keeping your beet patch weed free is vital to a good harvest. Weeds compete with any root crop, vastly diminishing yields. Start weeding when your seedlings are small, taking care not to damage or bruise young roots. Weed beets several more times throughout the growing season to minimize competition and maximize beet yields.

When the beet roots get to be about an inch in diameter, thin beets again to stand 8–12 inches apart. The small roots can be eaten fresh, or pickled whole. Now is the time to mound or hill the soil up around the growing roots. Water beets well during any dry periods. If beets receive less then an inch of water per week the plants will bolt, or go to seed, and the roots will crack and become tough and woody.

Fertilize beets once a week with manure tea for maximum growth and yield. Side dress beets with well composted manure once during the growing season. Immediately after hilling them is the usual time for side-dressing.

Insect pests and diseases

Well cultivated beets are usually disease and insect free, however, on occasion flea beetles or leaf miners will attack the leaves. These pests rarely do any serious damage to the plants.

Flea beetles are tiny, black beetles that jump like fleas when they are disturbed. The adult flea beetle chews small round holes in leaves, while larvae feed on plant roots. Control flea beetles by applying floating row covers, applying parasitic nematodes to the soil, or spraying with pyrethrums, rotenone or carbaryl.

Leaf miners are tiny black flies whose larvae tunnel through the plant's leaves. Control them by removing any foliage that is infected.

Boron deficiency can cause brown hearts and black spots to develop in beet roots as well as stunted growth. Rock phosphate or granite dust added to soil in early spring will prevent this deficiency. Foliar feeding of manure tea or seaweed extract will also help to minimize damage from the condition.

Harvesting beets and beet greens

You can begin harvesting beet greens when the plants are still small, but the tops are abundant. Continue to harvest throughout the growing season. You can harvest up to 1/3 of the plants greens without harming the plants.

Beets should be harvested when the roots are from 1 ½ to 3 inches in diameter. Leaving roots in the ground after they have reached maximum size will result in tough, woody beets. Hand pull beets carefully to avoid bruising them. Shake excess soil off, and remove tops about 1-inch from the top of the root to prevent bleeding of the beet root. Can or freeze the greens you cut off from the roots.

Beets can be kept for up to six months by layering fresh, undamaged roots in a box containing clean sand or sawdust, and storing in a cool place. Beet roots can also be frozen, canned, or pickled.

Broccoli

Broccoli can be cooked fresh, eaten raw, and frozen. Broccoli is rich in vitamins A and C, as well as the B vitamins, calcium, and iron.

Broccoli prefers full sun, but planting in partial shade can help prevent plants from bolting, or going to seed, in areas with hot, dry spells.

Broccoli is a cool weather crop that needs rich, well drained soil. Work in plenty of compost before planting. Cool days and nights are essential for broccoli once the heads begin to form.

Broccoli can take anywhere from 45 to 85 days to mature, depending upon the cultivar chosen. Try to select a cultivar that will mature before the weather in your area turns hot. In temperate areas you can harvest crops in both spring and fall.

Planting guidelines

Although it is possible to sow broccoli seed directly out in the garden in the spring or fall, temperature related deformities in head growth often occur. It is a much safer practice to sow broccoli seed indoors about 8 weeks before the last expected frost date in your area. Plant each tiny seed ¼ inch deep in it's own pot. Little peat-pellet pots are ideal to use. Place pots in a sunny area and maintain the temperature at around 65 degrees F. Be sure to keep the soil moist, but not wet. Broccoli seed should germinate in about 5 days.

For robust plant, and proper head production it is crucial to get your plants transplanted into the garden at the proper size. Plants should be about 6 inches tall, with 4 leaves. Started plants can also be purchased from your local garden center. Harden plants off by setting them outside during the day and bringing them back in at night. Do this for a week before transplanting the plants into the garden.

When transplanting broccoli, set the plants 1 or 2 inches deeper then they were setting in the pots. Space plants 1 to 2 feet apart in rows that are set 2 to 3 feet apart. Plants spaced too close together will develop smaller heads, so be sure to allow ample room between plants. Firm the soil up around the roots, and water in well. Fall crops can be sown directly in the ground.

Maintenance

Around two weeks after transplanting, top-dress your broccoli with manure tea, or side dress with well composted manure or fish emulsion, and water deeply. Repeat this process monthly until a week before harvesting the heads.

Keep broccoli plants cultivated to remove weeds and keep soil loose. Don't allow the soil to dry out too much or your plants will produce tough stems.

Insect pests and diseases

Possible broccoli pests include aphids, cabbage loopers, imported cabbageworms, cabbage maggots, cutworms and flea beetles, as well as slugs, mites, and harlequin bugs.

Slugs are obvious pests with their thick, usually tan-colored, muscular bodies that are covered with slime. Slugs often leave a slime trail behind them as they move about. Slugs are not insects but rather mollusks, and are related to snails. Slugs chew holes in plant leaves, and often attack fruits as well. Slugs range in size from less then 1 inch to over 4 inches in length. Control slugs by handpicking and destroying. If you set out rocks and boards to attract beneficial insects to your garden, you will also attract slugs. Use these items as slug traps and check under each board or rock daily, killing any slugs found hidden beneath them. You can also sink shallow containers of beer in the soil near your plants where slugs are a problem. The beer will attract the slugs which will climb in to drink, and subsequently drown.

Cabbage worms and cabbage loopers can be controlled by handpicking of insects or by applying carbaryl, rotenone or pyrethrums. Destroy any eggs you find. They will appear as tiny yellow dots on the undersides of broccoli leaves.

Diseases are seldom a problem with broccoli. Among some of the more common diseases are Black leg, black rot, club root, and fusarium wilt.

Black leg produces dark spots on the leaves and stems of plants.

Black rot causes yellowing leaves with dark, foul smelling veins. Prevent both of these diseases with good cultivation and crop rotation.

Club root is a dieses that causes yellowed, weak, stunted plants with deformed roots. Club root can be eliminated by the addition of lime to the soil, adjusting pH to 7.0.

Fusarium wilt, also known as yellows, causes lower leaves to turn yellow and drop off and makes broccoli heads stunted and bitter tasting. Destroy any afflicted plants to prevent spreading of this disease.

Harvesting broccoli

You should watch your broccoli closely during the development of flower heads. Harvest the heads before the flower buds open and turn yellow. Cut the head off at a point just below where the main stem splits into multiple stems. Once the main head has been harvested, smaller but equally delicious side heads will form. Keep picking these side heads and your broccoli will continue to produce until the weather gets too hot or too cold.

Although best used fresh, broccoli will keep, refrigerated, for up to 2 weeks. Broccoli can be frozen, pickled or eaten fresh, cooked or raw. To drive any hidden cabbage loopers or cabbageworms out of your flower heads, soak them in warm water with a little vinegar added for about 15 minutes.

Brussels sprouts

These miniature cabbages take up to 100 days to mature, however, they yield a large harvest from a very small area.

Brussels sprouts, like their relatives, are a cool weather crop, and are also the hardiest of the cabbage-family crops.

Planting guidelines

For a spring crop, start seeds indoors about 8 weeks before the last expected spring frost in your area. Sow seed ½ inch deep in individual pots. Peat pellet type pots are ideal for starting Brussels sprouts. Started plants can also be purchased from your local garden center. When plants are about 6 inches tall, transplant into the garden spacing plants 2 feet apart. Set the transplants deeper in the soil then they sat in the pots, firming the soil around the roots. Water plants in well.

Fall planted crops often yield heavier harvests then do spring planted crops. In areas where fall crops are possible, sow seed directly in the garden ½ inch deep and about 2 inch apart. When plants are about 6 inches tall, thin to stand about 2 feet apart.

Maintenance

Mulch Brussels sprouts well to retain soil moisture, and cultivate to reduce weeds. Use care when cultivating close to the plants not to damage growing sprouts on the plant stem, or bruise the shallow root systems. Brussels sprouts benefit from being staked in windy areas.

The leaves growing below each sprout will turn yellow as the sprouts mature. Pick off these yellowed leaves to give the sprouts more room to grow.

Insect pests and diseases

Brussels sprouts suffer from the same insect and disease problems as do cabbage. The worst of the insect pests to affect Brussels sprouts are cabbage loopers, cabbage maggots, imported cabbageworms, cutworms, slugs and snails.

Occasionally, harlequin bugs will infest Brussels sprouts causing leaves to develop black spots and wilt. Harlequin bugs are medium-sized black insects with red markings. Hand pick harlequin bugs you can see, or use an insecticidal soap.

Cabbage worms and cabbage loopers are both green caterpillars, the cabbage looper having a humpbacked appearance. They both chew large holes in the leaves of plants, occasionally reducing the leaves to skeletons. Control both pests by handpicking or by using an insecticide such as carbaryl as a preventative. The caterpillars can often be found on the insides of leaves along the leaf veins. If you are handpicking alone, be sure to crush and remove the tiny, yellow eggs that are deposited on the undersides of leaves.

Diseases are seldom a problem with Brussels sprouts. Black leg, a fungal disease, causes large dark spots to form on the leaves and stems.

Black rot is another disease that causes leaf veins to eventually become black and foul smelling.

Fusarium wilt, or yellows, causes leaves to turn yellow and heads to be stunted.

Club root is a root disease that prevents the plant roots from taking in water and nutrients from the soil, resulting in small, weak plants and deformed roots.

If any of your Brussels sprouts are affected by these diseases, remove and burn the affected plants to prevent the disease from spreading. Adjust the soil pH by raking a little extra lime into the soil to help deter club root once it has already appeared. Good sanitation, crop rotation, and the use of disease resistant cultivars are the best defense against these diseases.

Harvesting Brussels sprouts

Harvest Brussels sprouts as they mature from the bottom of the main stalk upward. Smaller sprouts are the most tender and delicious. Remove sprouts from the stalk by twisting them. To force sprouts to mature faster, pinch off the main plant's top.

Before the weather freezes in the fall, or before a summer heat spell you can pull the plants out of the ground and hang them upside down in a cool place to mature smaller sprouts higher up on the main stalk. This will prolong your Brussels sprouts harvest by a couple of weeks.

Cabbage

Cabbage comes in several different cultivars, from the common green heads we all know, to the slightly sweet red cabbage, loose-leaf varieties such as bok choy, and the ornamental cabbages. Cabbages heads can weigh anywhere from 2 to 50 lbs, depending on the cultivar and growing conditions.

Planting guidelines

Plant cabbage seed about four weeks before the average last spring frost date for your area. Sow seed indoors, ¼ inch deep in individual pots, two seeds to a pot. Pellet type peat pots are excellent for this use. Place pots in a sunny spot where temperatures will remain between 60 and 70 degrees F. Keep the soil uniformly moist.

When plants emerge thin to the strongest plant per pot. Started plants can also be purchased from your local garden center.

When outside daytime temperatures reach 50 degrees F and the seedlings have three leaves, plant them outside in the garden bed.

Plant seedlings in the garden slightly deeper then they grew in the pots. Space plants 6 to 12 inches apart, depending upon cultivar size. Keep rows 1 or 2 feet apart. Wide spacing allows for larger heads that store well, however, younger, smaller heads are often the tastiest. To get both size heads, plant cabbages 6 inches apart and harvest every other one before they reach full maturity. Make planting holes wide enough to accommodate spread-out roots. Firm soil around plants and water thoroughly.

Start your late crop in July. Plant seed directly out in the garden. Space these seedlings farther apart than the spring crop, and place them where a taller crop, such as corn or pole beans, can provide some afternoon shade and relief from the heat of the day.

Maintenance

Side-dress your cabbage seedlings with well-composted manure three weeks after planting, and again mid-season. Hand-pull weeds to avoid damaging cabbages shallow root system.

It's always good to use a mulch to keep the soil moist around cabbage plants. Uneven watering can cause a sudden growth spurt that will make the heads split. A really rainy spell can do the same thing. If you see a head that is starting to split, twist the plant a half turn and pull up to slight dislodge the roots and slow the plant's growth. You can also use a spade to cut the plant's roots in one or two places about 6-inches below the stem. Cutting the roots in this manner also helps to prevent the cabbage plants from bolting, or forming flower stalks.

Insect pests and diseases

Cabbage loopers, imported cabbage worm, cabbage maggots and cutworms are the most common insect pests of cabbage plants.

Cabbage loopers and cabbage worms are both thick green caterpillars, the cabbage looper having a distinctive hump in it's middle. Hand pick any of these pests when you see them. Cabbage worms and cabbage loopers like to hide along the leaf veins, blending in amazingly well with the plant. The eggs of these insects are tiny yellow and whitish eggs that are laid on the underside of the leaves. These insects also prefer to feed on the more tender and younger, inner leaves of the plants, although on occasion they will feed on the tougher, older leaves on the outside. Carbaryl and insecticidal soaps also work well to control these pests.

On occasion slugs will infest cabbages. Slugs are thick, muscular, tan colored mollusks, related to snails. Hand pick slugs whenever you see them. Shallow containers of beer set in the cabbage patch will attract and drown these pests. Stones or boards set out around the garden will also attract slugs. Be sure to check beneath these traps daily and destroy any slugs found hidden beneath them.

Black leg, a fungal disease, causes dark spots to form on leaves and stems. Black rot causes leaf veins to become black and foul smelling.

Club root causes misshapen roots and stunted plants due to hampered water and nutrient absorption through the damaged roots.

Fusarium wilt, also known as yellows, produces yellow leaves and stunted heads. Remove all plants affected by any of these diseases and destroy them to prevent the spread of the disease. Good sanitation, crop rotation and good growing conditions will all help to prevent these diseases. Always buy disease resistant cultivars for the best defense against disease.

Harvesting cabbages

Cut heads with a sharp knife when they are firm. Leave the stalks in the ground to produce tasty little cabbages. Eat them like Brussels sprouts or let them develop into a second crop of smaller cabbages. Although fresh cabbage has the best flavor, late season cultivars can be stored in a cool, moist place for up to 6 months. Split heads can be turned into excellent sauerkraut.

Carrot

Carrots come in a wide assortment of colors and shapes for the home gardener to try. There are white, yellow, orange and crimson roots that can be fat, slender, or round. There are late carrots, and early carrots, as well as disease and crack resistant cultivars.

A single carrot every day provides more then the recommended dietary allowance of vitamin A, as well as supplying needed vitamins B, C, D, E, and K.

Planting guidelines

Carrots demand rich, deep, loose soil. It is vital that the soil be rock and clod free if you are to have a truly successful crop.

To produce the best crop possible, prepare your carrot bed by double digging or tilling the soil until it is smooth and free from all dirt clods.

Well composted manure is a good addition to the carrot bed, but try to avoid any fresh manure as the nitrogen will produce poor-tasting, branched carrots.

Carrots are a cool weather crop, so plan to sow the seed directly into the garden about three weeks before the last expected spring frost in your area. Plant again about 3 weeks later. Carrots can be sown in the fall in subtropical regions.

Carrot seeds are tiny, so the easiest way to sow them is to sprinkle a pinch of about six seeds over an inch of row. Cover with ¼ to ½ inch of soil and firm well. Carrots take between 1 and 3 weeks to germinate. Some homesteaders prefer to plant fast growing radish seeds in amongst the carrot seed to mark the row and loosen the soil for the slower growing carrot seedlings. Water seed in gently to avoid washing any away. Keep the soil moist so the seed will germinate quickly.

Maintenance

When the carrot tops are about 2 inches high, thin to stand 1 inch apart. Be ruthless in thinning because crowded carrots will usually be deformed. Thin again two weeks later to stand 3 to 4 inches apart.

A layer of mulch will help the carrot bed to stay evenly moist. If your carrot bed dries out between waterings, gradually add water over a couple of days. A sudden drenching can cause carrot roots to split.

Cultivate and remove weeds by hand to prevent damaging the tender roots. Cover crowns as they appear throughout the growing season to prevent carrots from becoming green and bitter tasting.

Insect pests and disease

Mammals such as deer, woodchucks, rabbits and gophers are the biggest threat to carrots. Garden fencing, traps and the liberal sprinkling of human and dog hair around your garden will help to deter these animal pests.

In the Northwest especially, carrot rust flies can be a problem. They appear to be tiny green houseflies, about ¼ inch long with yellow heads and red eyes. Their eggs hatch into whitish maggots that burrow into roots causing the carrots to turn dark red and the leaves the to turn black. Covering the plants with floating row covers to keep the flies away is usually successful. Because rust fly infestations usually occur in the early spring, it is often beneficial to delay planting carrot seed until early summer where weather permits.

Parsley worms, green caterpillars with black stripes and white or yellow dots, attack carrot foliage. Hand pick these pests or apply rotenone, or carbaryl for severe infestations.

Carrot weevil larvae tunnel into carrot roots, especially spring crops. Crop rotation is the best remedy for this pest.

Nematodes, microscopic wormlike animals, make little knots along roots that result in distorted or stunted carrots. Crop rotation and plenty of compost discourage nematodes. For severe infestations, plant your carrot bed with French marigolds the year before planting carrots there.

Leaf blight is the most widespread carrot disease. Leaf margins develop white or yellow spots which turn brown and watery. Purchase leaf blight resistant cultivars to avoid this disease.

Vegetable soft rot is a bacterial disease that occurs during hot, humid weather. Prevent it by rotating crops, and keeping the soil loose. Do not store bruised carrots as the disease is spread in storage.

Mosaic virus mottles leaves with light to dark shades of green. Aphids spread the virus so controlling aphids in your carrots will, in turn, control this disease.

Harvesting carrots

You can harvest baby carrots as soon as they are big enough to eat, or you can wait until roots are mature. Larger, more mature carrots usually have better flavor. Dig your winter storage crop before the first fall frost is forecast. The soil should be moist, but the air dry. Hand pull the roots to prevent damage which could lead to spoilage in storage. If the soil is dry, water the morning of harvesting to loosen the soil and make pulling the carrots easier.

To store carrots for winter use, twist off the tops and place a layer of carrots in a box that has a thin layer of sand on the bottom of it. Put a layer of sand over the carrots and lay another row of carrots atop that. After a final layer of sand, place a top layer of peat or straw on the box and place the box in a cool place. Storing carrots in this manner preserves more of their flavor and nutritional value then does canning or freezing.

Cauliflower

A mild-flavored member of the cabbage family, cauliflower is not difficult to grow, however, it is very sensitive to temperature extremes. Like other cabbage family members, cauliflower is a cool weather crop. It will not produce the delicious flower heads in hot weather, and is only frost tolerant as a mature fall crop. It is important for maximum success with your cauliflower crop that you choose a cultivar that is correct for your growing area and that you follow strict planting and growing guidelines, and supply ample moisture throughout the growing season.

Planting guidelines

Like other cabbage family members, cauliflower needs a rich, well tilled garden soil. Although usually planted in the spring by homestead gardeners, cauliflower works best as a fall or winter crop in most areas.

For a summer crop, plant seeds indoors at least 4 weeks before the last expected spring frost for your area. Sow seeds ¼ to ½ inch deep in individual pots. Pellet type peat pots are ideal for this purpose. Set the pots in a south-facing window and be sure to provide constant moisture. Use bottom heat, such as a heating pad, if necessary, to keep the soil temperature at or about 70 degrees F.

Cauliflower plants can also be purchased from your local garden center.

About two weeks before the expected last spring frost date in your area, harden off the plants by setting them outside during the day and bringing them indoors at night. Examine your seedlings to be sure each one has a small developing bud in the center. Any plants without this bud should be disposed of because they will never go on to produce a head.

Transplant seedlings into the garden after the last frost date, but while weather is still cool. Plant seedlings 15 to 24 inches apart in rows set 2 to 3 feet apart. Firm soil around roots and water thoroughly.

For a fall or winter crop, sow seed directly out into the garden bed 2 to 3 months before first expected fall frost in your area. Sow seed in hills of four seeds per hill, and water thoroughly. Approximately one month later, thin seedlings to the strongest plant per hill.

Maintenance

Cauliflower requires constant moisture to produce large, tender heads. Soil that dries out between waterings will cause heads to open up and develop a ricey texture. A thick layer of compost and mulch will help preserve moisture and cut down on evaporation. It also helps keep the soil cooler in hot weather.

Weed and cultivate around cauliflower by hand so as not to damage the shallow roots. Give your young cauliflower monthly or bi-weekly feedings with manure tea.

When the flower heads of white cauliflower cultivars are about the size of an egg, you should blanch them by shading out the sunlight. If you neglect this step you will end up with green or brown heads that are not very appealing. Choose a sunny afternoon when the plants are completely dry. Damp cauliflower heads at blanching time will increase the likelihood of the heads rotting. To blanch cauliflower just bend some of the plants' own leaves over the head and secure them with soft twine, rubber bands or plastic tape. Use enough leaves to block out light and moisture while still leaving room for air circulation and for the heads to grow.

Once the blanching process is underway, water only the roots of the cauliflower plants, not the heads or leaves. Unwrap your plants occasionally to check for insects, disease, or to allow the heads to dry out after a rain. In hot weather, heads can be ready to harvest within days of blanching. In cooler weather the process can take up to two weeks. This blanching process is not necessary with self-blanching varieties or with the purple headed varieties.

Insect pests and diseases

The same insect pests that attack other members of the cabbage family also cause serious problems in cauliflower. Aphids and flea beetles tend to bother cauliflower more in the spring then in the fall. Cabbage maggots or root maggots can be a serious problem.

Cabbage loopers and imported cabbageworms are common cauliflower pests. Control these green, fleshy caterpillars by hand picking, or apply rotenone, or carbaryl to your plants. Harlequin bugs, black insects with red markings are another cabbage family pest. Control them by handpicking, or use insecticidal soaps, rotenone or carbaryl.

In areas with boron deficiency, cauliflower heads turn brown and leaf tips die back and become distorted. If this deficiency strikes your plants, foliar-feed with liquid seaweed extracts immediately, and repeat the feeding ever two weeks until the symptoms disappear. For subsequent crops in boron deficient soil, add granite dust or rock phosphate to the soil. A cover crop of vetch or clover sown during off-season will also help eliminate this condition in the soil.

As with other cabbage family members, black rot, club root and fusarium wilt are all fairly common diseases of cauliflower.

Black rot produces yellowing leaves and black, foul-smelling veins.

Fusarium wilt causes leaves to turn yellow and flower heads to be stunted. Club root causes roots to become malformed and interferes with moisture and nutrient absorption. Remove and destroy any plants affected by these diseases. As with most diseases, good sanitation habits, crop rotation, and choosing disease resistant cultivars will help eliminate the potential of disease outbreaks.

Harvesting cauliflower

Mature flower heads can range in size from 6 inches to 12 inches in diameter. Harvest heads when the buds are still tightly closed. Using a sharp knife, cut the heads off just below the base of each head. Be sure to include a few of the closest leaves to protect the head. Cauliflower can still be used if it has been hit by a heavy frost. You just cannot re-freeze cauliflower that has been hit by a heavy frost.

Cauliflower can be frozen, pickled or eaten fresh. To store fresh cauliflower for about a month, pull up the entire plant, roots and all, and hang upside down in a cool place.

Celery

Celery is a fairly easy plant to grow. It

does require rich soil, plenty of moisture, and protection from high temperatures and hot sun. Celery can be grown as a spring crop in cooler climates, and a fall or winter crop in warmer climates.

Planting guidelines

Plant celery seed indoors 8 to 10 weeks before the last expected spring frost in your area. Soak the tiny seeds overnight to encourage germination. Fill flats with a mixture of 2/3 compost and 1/3 sand. Plant the soaked seed in rows 1 inch apart. Cover the seeds with a sand layer 1/8 inch deep, then cover the flats with a layer of damp sphagnum or burlap until the seeds germinate.

Place the flats in a bright spot away from direct sunlight. Keep temperature around 75 degrees F. during the day, and around 60 degrees F at night. Provide the flats with plenty of water and air circulation.

When seedlings are about 4 inches tall, transplant them to individual peat pots. At 6 inches tall, harden off plants by putting them outside during the day and bringing them back in at night for about ten days. After this time, transplant them into the garden bed.

Space celery plants 6 to 8 inches apart in rows spaced 2 to 3 feet apart. Set them no deeper than they grew in the pots. Water them immediately upon transplanting, and fertilize each seedling with manure tea.

For a fall crop, sow seeds indoors in May or June, and follow the same basic planting instructions, transplanting seedlings in July. Provide the plants with shade during hot, humid weather.

Maintenance

Apply several inches of mulch around the base of plants. Keep plants evenly watered throughout the growing season. Hand cultivate and weed to remove any weeds that may compete with celery's shallow roots for water and nutrients. Fertilize every two weeks.

Blanching prevents celery stalks from becoming bitter, and adds a certain amount of protection from the weather. Some nutrients will be lost due to blanching, however, most homesteaders feel that the tradeoff is worth it. Blanching can be accomplished by any of several means. You can begin blanching early, while the plants are young and then continue as they grow. Gradually pull the soil up around the plants as they grow throughout the season. Keep only the leaves exposed. You can also wait until two weeks before harvest, and then tie the tops together, mounding the soil up to the base of the leaves. Or try covering the stalks with large cans, drain tiles, or sleeves made out of paper or other material. Any of these methods will result in satisfactorily blanched celery.

Insect pests and diseases

Celery's main enemies are parsley worms, which can defoliate celery in a matter of hours, carrot rust flies which appear to be tiny green houseflies, about ¼ inch long with yellow heads and red eyes, and nematodes. Celery leaf tiers are tiny yellow caterpillars marked with one white stripe. Control parsley worms, carrot rust flies, and celery leaf tiers by handpicking or with the application of carbaryl or pyrethrims.

Early and late blight both affect celery crops. Both begin as small dots spreading on the leaves.

Pink rot is another disease affecting celery. It shows up as water soaked stem spots, and white or pink discoloration at stalk bases.

Crop rotation and good garden hygiene are the best courses of treatment for these diseases. Distorted leaves and cracked stems can indicate a boron deficiency in the soil. Correct the deficiency by spraying plants with a liquid seaweed extract every two weeks until symptoms disappear.

Harvesting celery

Cut the entire blanched plant off just below the soil line. You can also cut single stalks of unblanched celery as needed.

To preserve a fall crop, pull up entire plants, roots and all, and place them upright in deep boxes with moist sand or soil packed around the roots. Store these sand boxes in a cool place and the celery will keep for several months.

Chard, mustard greens and spinach

Swiss chard is a prolific plant that is grown for its large, mild flavored, succulent leaves. Chard withstands temperature extremes and changes better then most other greens. Chard tolerates partial shade as well as direct sun, although in partial shade the yield will be smaller.

Chard, mustard greens, and spinach will all tolerate a wide range of soil types. The only thing that is a threat to these plants is low soil pH, which results in stunted plants. Soil pH between 6.0 and 7.0 is ideal for all varieties of pot greens.

Planting guidelines

Chard seeds are actually a cluster of several seeds that will produce more than one plant. The spacing between seeds will determine size of the plants. Broadcast chard seed directly in the garden 2 to 3 weeks before the last expected spring frost in your area. Sow at a rate of one seed every 2 inches. Rake lightly to cover seeds. Thin young plants to stand 4 to 6 inches apart. Chard can also be planted in the late summer for a fall crop.

Spinach should be sown in furrows that are about ½ inch deep. The rows should be spaced about 1 foot apart. Spinach matures quickly and bolts easily, so plant in short rows and make succession plantings every 10 days until daytime temperatures reach a constant 70 degrees F.

Mustard greens should be sown early in the spring in northern areas, and in early fall in more southerly climates. Sow mustard seeds 1 inch apart in rows that are spaced about 1 to 1 ½ feet apart. Cover seeds with ½ inch of soil.

Maintenance

Water during dry spells, and remember, too much water early in the seedling stage can result in damping-off and other seedling disorders. Cultivate to keep the beds weed-free. Thin chard and spinach gradually until plants are 8 to 11 inches apart. Mustard greens should be thinned to stand 15 inches apart. Young chard and spinach plants thinned out make excellent additions to the salad bowl, and mustard greens can be steamed at any age.

Insect pests and diseases

Chard is an easy care plant that is relatively free from diseases. Remember not to cultivate during wet periods to help prevent the spread of any disease that may be present.

Insects that attack chard are aphids, leaf miners, and slugs. Control aphids and leaf miners with the application of rotenone or carbaryl. Remove and destroy any leaves infested with leaf miners.

Hand pick slugs by using a flashlight at about dusk. Boards and stones can be laid around the slug-prone area of the garden, and early in the morning the slugs can be collected from beneath these hiding places and destroyed. Slug traps can be fashioned using a coffee can filled with beer. Slugs are attracted to the scent of the hops or yeast, and they drown in the beer. If you have raised beds that are infested with slugs, simply stapling a copper strip around the border of the beds will deter slugs. Copper reacts with the slugs body to produce a sort of electrical shock that the slugs subsequently avoid.

Harvesting greens

Young chard and spinach leaves can be picked individually and used in salads and sandwiches. When harvesting older plants, pull the entire plant. Shake off excess soil from roots. Pick off and discard any leaves that are insect eaten. Plunge plant or leaves in very cold water for 10 to 15 minutes. Dry on paper towels and store in refrigerator until ready to use.

Harvest mustard greens before the plants are full grown or the plants may go to seed. You can harvest leaves for use as needed, or you can pull the entire plant out and harvest all the leaves.

Chard spinach and mustard greens can all be frozen or canned for future use.

Collards and Kale

Collards and Kale are closely related and are both members of the cabbage family. They are strong flavored, leafy vegetables that are relished in southern states. Like most members of the cabbage family, collards and kale prefer cooler weather. Hot weather creates bitter flavors in these vegetables.

Neither collards nor kale form the large, round heads that cabbage does. Some varieties of Kale are ornamental and are used in flower gardens.

There are two varieties of kale available. One is the curly-leaf type, and the other is the plain, or straight, leafed type.

Planting guidelines

Warmer areas can grow both a spring and a fall crop. All members of the cabbage family can withstand some frosts so don't feel that you have to harvest everything before frost hits in the fall.

Start your seeds indoors four to six weeks before last expected spring frost. Use peat pots to minimize handling of the seedlings when transplanting. Sow a couple of seeds ½ inch deep in each peat pot. Water well, and set in a sunny window to germinate.

When seedlings emerge, thin to the strongest plant in each peat pot. Transplant your seedlings outdoors as soon as the ground can be worked in the spring. Space seedlings about 12 inches apart in rows spaced about 2 feet apart.

Collards and kale can also be direct seeded directly in the garden bed, although with the transplanting method you can exercise more control over the spacing of the mature plants. To seed directly into the garden, sow seeds from 6 to 8 inches apart in rows that are about 2 feet apart. When seedlings emerge, thin seedlings to stand 12 inches apart.

Maintenance

Occasional side dressing with well composted manure will help to ensure a good yield of collard greens or kale.

Don't allow the ground to dry out between waterings, but do not allow the soil to remain soaked. A layer of straw mulch laid down between rows will help to maintain soil moisture and reduce weed problems.

Insect pests and diseases

All members of the cabbage family are extremely susceptible to insects. Collards and Kale are no exception. Among the most common insect pests to attack collards and kale are aphids, and cabbage loopers.

Cabbage loopers the larval stage of a medium sized white moth. Sometimes you can see the moths hovering and fluttering over your cabbages, collards, and broccoli plants. Effective treatment involves place a mesh screen over the plant so the moth can not lay her eggs. Carbaryl and rotenone are also very effective in controlling this pest.

Organic controls include hand picking the tiny yellow eggs that can be found cemented to the underside of leaves. Insecticidal soaps, as well as garlic and hot pepper sprays are also effective methods of control.

Aphids are also controlled by the use of Carbaryl and rotenone. For organic control measures, insecticidal soap can be used as well as a hard stream of water aimed at the plants to wash off the insects. If this route is chosen, you will have to repeat it on a weekly basis.

Very few diseases affect collards and kale.

Harvesting
The young leaves can be harvested as the plant grows for use in salads, soups and other recipes.

Continual harvesting is important in keeping your plants productive and flavorful. Letting leaves remain to get over-ripe will result in tough, bitter leaves.

When the weather heats up and plants get too big, and the outer leaves start getting tough, you can harvest the entire plant, stripping away the older, outer leaves, and using the tender inner rosette of leaves.

Corn; sweet, popping, and dent
These basic guidelines are for planting and growing sweet corn, although popping corn and dent corn, or field corn, are grown in the same manner.

Even in a small garden, sweet corn and popping corn can be grown with success. The biggest reason for corn crop failure in the homestead garden is pollination problems, which can easily be overcome with proper planting and spacing. Poor pollination results in poorly filled, small ears.

Corn is wind pollinated with the actual pollen being produced by the tassel part of the plant. The pollen is then blown by the wind onto the silk portion of a plant. When the corn is spaced too far apart or in only one or two long rows, more pollen is blown away then is actually landing on the silk, therefore, poorly filled ears develop. Any strand of silk that does not get a particle of pollen land on it will never develop into a kernel of corn. There will just be a gap on the ear where that kernel would have been. Always remember, for pollination purposes, it is better to plant corn in several shorter rows then one or two longer rows.

Corn is a warm season crop that requires a minimum soil temperature of 50 degrees F for soil germination. All corn varieties need full sun for maximum growth and yield.

Sweet corn requires rich soil with ample nitrogen and moisture. Amend the soil with well-aged manure or compost. If you plant corn in an area that had healthy beans or peas grown in it the previous year it is helpful because these legumes contribute more nitrogen to the soil.

Because corn plants grow so tall and spindly they are very susceptible to wind damage. The chance for wind damage is reduced by planting corn rows in blocks of at least 4 rows for support. Planting in blocks of rows also helps ensure the best pollination rate.

"Normal" sweet corn (su) is corn with kernels that contain moderate but varying levels of sugar, depending on variety. Sugars convert to starches rapidly after harvest. All of the open-pollinated heirloom varieties are normal sweet corn.

"Sugar-enhanced" (se, se+, or EH) is corn with a gene that increases tenderness and sweetness. Additionally, conversion of sugar to starch is slowed.

"Super-sweet" or "Xtra-sweet" (sh2) is corn that greatly increases sweetness and slows the conversion of sugars to starch. The dry kernels (seeds) of this type are much smaller and more shriveled.

Planting guidelines

Corn should be planted no earlier than two weeks after the last frost in the spring. Ideally, the soil temperature should be between 60 and 80 degrees F for the best germination rate.

Corn plants should be spaced at 8"-10" for early cultivars and 9"-12" for late cultivars, with rows spaced 30"-36" apart.

If you plan on saving seeds and want to plant more than one cultivar of corn, space the cultivars at least 250' apart to prevent cross-pollination. An alternate method to prevent cross-pollination of different corn cultivars is to stagger planting dates of the different cultivars by a minimum of 14 days. Naturally you will want to separate your popping, dent and sweet corn from one another to prevent any confusion.

Sow corn seeds by planting one seed every 3 inches, about 1" deep, within rows spaced 30"-36" apart. Gently tamp soil in bed after planting seed. Thin corn seedlings to stand 6 to 8 inches apart.

For a continuous supply of sweet corn throughout the growing season, include early, mid-season, and late cultivars in your initial planting. You can also make 2 or more successive plantings of each cultivar at 2 week intervals to extend your sweet corn harvest.

Because of its fast germination as well as plant tenderness, corn is not normally transplanted.

If for some reason you want to get a

head start on the season and decide to start your sweet corn indoors, be sure and plant it in small peat pots that can be planted directly in the ground, thereby minimizing handling of the tender plants.

Germination will take 7–10 days.

Maintenance

Proper watering is vital to good ear formation, so it is important to maintain an adequate level of soil moisture during two critical periods of corn development. Corn water consumption will generally rise dramatically as the plants approach both tassel and silk stages.

A properly spaced block of corn uses water quite efficiently. Corn that is poorly spaced will experience excessive water loss from soil exposure.

As corn grows it is often helpful to hill up the corn similar to the way you would hill up potatoes. Using a hoe, gently pull soil up from between rows to heap around the root-line of the plants. This will help conserve moisture around the roots as well as adding important support to the growing plants. Be careful in cultivating close to the plants that you do not disturb the roots themselves.

Sweet corn sometimes has a tendency to develop side stalks, or suckers. Remove these suckers when you find them, leaving only the sturdy main stalk growing. Leaving the suckers will not increase yield, rather, they will draw so much energy from the plant that the yield will be reduced.

Insect pests and diseases
Corn Earworm, corn borer, flea beetles, and cutworms are some of the most common pests of corn. Treat with pyrethrins, rotenone or carbaryl, especially when in tassel and silk stages. A nice, natural control method for corn earworms is to pour a few drops of mineral oil onto the silks about ten days after the silk appears.

The best treatment for corn borers is prevention. Corn borers bore into corn stalks and will over-winter in them. Because of this, it is very important to clean up and remove all cornstalks and corn debris from your garden at the end of each growing season. Corn stalks can be fed to livestock, chipped up for animal bedding or compost, or simply burned if borers were too great of a problem that year.

Corn smut is a fungal disease that can appear on the stalks, leaves, tassels, or the ears of sweet corn, popping corn and dent corn. Dent corn is somewhat more resistant to the disease than is sweet corn or popping corn. Later cultivars of sweet or popping corn are also somewhat more resistant then are the earlier varieties. Corn smut is most obvious when it attacks the ears of the plant because each kernel fills with smut fungus spores turning the kernels into purplish black globs that are visually active along the ears. There is no effective treatment for corn smut, so the best defense is a good offense.

Good sanitation practices in the garden will help to prevent the fungus from getting a foothold to begin with. Do not feed corn stalks that have been infected with corn smut to livestock if the manure from those animals will be used on the garden in the coming months because the corn smut spores will still be viable and can be spread in the garden from that infected manure. Corn smut can also be spread as spores by the wind from a neighboring field that is infected with the disease.

Use care when cultivating around plants not to damage the plants or roots, thereby giving any spores easy entry into your plants. Rotating your corn crop will also give some degree of protection, Bacterial wilt, or bacterial leaf blight are two common names for Stewart's disease of corn. Stewart's disease is more destructive to sweet corn than it is to dent or popping corn, however, some of the hybrid strains of both field and popping corn are quite susceptible to the disease.

The corn flea beetle is the sole source of the spread of this destructive disease. Symptoms of Stewarts disease usually appear as leaf lesions resulting from feeding scars of corn flea beetles. As the disease spreads within the plant, entire leaves wither and die and the plants become stunted and brown. The entire plant may wilt and die.

Again, the best control is prevention by planting disease resistant cultivars. If corn flea beetles, or Stewart's disease are a problem in your area, you will probably want to buy pre-treated corn seed. This pretreatment is good until the corn plants have about 5 leaves per plant. The application of carbaryl will help to control active flea beetles on growing and mature corn plants.

Harvesting

Sweet corn is fully mature when the stalks are anywhere from 5'-7' tall or more and have at least one or two ears. The corn silk should start to turn brown and slightly dried on the edges, and the kernels should be full to the touch and yield a milky white fluid when broken with a thumbnail.

To harvest all corn cultivars, pull down quickly on the ear of corn, twisting it at the same time to separate it from the stalk.

Sweet corn will quickly begin to convert sugars into starch after harvest so it is very important to quickly cool the ears after harvest. The rate of conversion of sugars to starch in the corn kernels increases with the rise of temperature. Sweet corn should be stored at close to freezing with a relative humidity of 98%-100%. In ideal conditions sweet corn may last up to 4–6 days without becoming overly starchy.

Leave popcorn in the garden until the stalks and husks are all brown and dry. When you can no longer leave a mark on the kernel with your fingernail, the popcorn is ready to harvest. Twist and snap each ear from the stalk. Do this before the first frost hits in the fall. Once harvested from the garden, popcorn needs time to cure or completely dry before being stored.

To prepare popcorn for indoor curing, carefully strip away the dried husk from each ear. The kernels will be only partially dried at this point, even though they may feel completely dried. It will take another 4 to 6 weeks of thorough drying in a warm, well ventilated place before the kernels can be stripped and stored.

Place the ears in mesh bags or spread them out on a large screen in an area where they'll have warm air circulating around them. You can also hang the mesh bags full of popcorn ears in your garage for about four weeks.

After curing, rub the popcorn kernels off the ears by hand, or you can leave the kernels on the ears and hang the bags of corn in a cool, dry place. The popcorn can keep for years in the cool, dry, dark conditions.

Dent corn is harvested in the same manner as popcorn. When the kernels are completely dried they can be stored whole, or immediately ground into cornmeal and stored in an airtight container in a cool, dry, dark place.

Cucumbers

Cucumbers belong to the cucurbit family, which also includes squashes, pumpkins, and melons. Cucumbers are easy to grow and tend to be prolific producers. They can be grown either for pickling or slicing. Although cucumbers require substantial growing space, they can be grown on trellises or other vertical structures to conserve garden space. Cucumbers can also be grown quite successfully in containers.

The cucumber ranges in size from the small gherkin type to the long, thin slicing variety. Cucumbers also come in both green and yellow color variations.

Although originally a sub-tropical plant there are now many new varieties that require a shorter growing season making them ideal for homesteaders in more northern areas.

Cucumber vines bear both male and female flowers on the same plant. The first flowers are male and will drop from the vine without producing any fruit. Subsequent flowers will include both male and female, and will go on to produce fruit. Recently, gynoecious plants (those bearing female flowers only) have been introduced to our markets. These seed packets will have specifically marked seeds that must be planted as well for proper pollination. The yield of these plants is somewhat greater than that of other cucumber vines.

Planting guidelines

Cucumbers can be grown successfully in many types of soils. They prefer rich, well drained soil. The ideal soil pH should be between 6.0 and 7.0.

Cucumbers can be planted indoors a couple of weeks before planting outdoors. Plant 2 seeds per peat pot or pellet. Water well and set in a sunny window. Transplant outdoors after all danger of frost has passed in the spring. Transplant by planting the entire peat pot into a small hole dug into the soil. Cucumbers can also be seeded directly out in the garden after all danger of frost has passed in the spring. Cucumbers can be planted in rows for easy trellising or they can be grown in hills or mounds, the vines being allowed to trail down the mounds. If planting in rows, plant 2 plants or seeds every 3 feet apart with rows spaced 6 feet apart. If planting in mounds or hills, plant four plants or seeds per hill, with hills spaced about 4 feet apart.

Maintenance

Water cucumber plants during any dry spells. Do not over fertilize as this encourages vine growth and retards fruiting. Mulching will conserve soil moisture, prevent soil compaction, and will help prevent the fruits from rotting where they touch the ground. Mulching also helps suppress weeds.

If staking or trellising cucumbers, the trellis, fence, or other support should be place solidly in the ground before the plants or seeds are placed in the garden. The vines are gently placed and trained up the support. Vines grow quickly and should be checked and directed along the support on a daily basis.

Insect pests and diseases

Striped cucumber beetles, aphids, mites, and pickle worms are the common pests of cucumber plants. Carbaryl, insecticidal soaps, and pyrethrins are all useful in controlling most of these pests. It is especially important to control cucumber beetle outbreaks because the beetle is the usual culprit in the spread of bacterial wilt as well as mosaic virus in cucumbers.

Bacterial wilt, mosaic virus, angular leaf spot, anthracnose, blossom blight, and powdery and downy mildews, are all potential disease problems in cucumbers.

Cucumbers can be infected with bacterial wilt by cucumber beetles when seedlings are just emerging from the ground. Bacterial wilt causes plants to wilt and die. The treatment for bacterial wilt is to destroy any infected plants, and to implement some form of control over the cucumber beetle.

Mosaic virus is also spread by the cucumber beetle. Control of this beetle is vital in the control of these two important cucurbit diseases.

Angular leaf spot causes leaves to develop square to rectangular water-soaked spots that turn light brown to white. Angular leaf spot can be controlled by spraying with a copper based fungicide.

Anthracnose can be particularly severe when the weather is wet and the temperatures are 70 to 80 degrees. Symptoms of anthracnose include round, water-soaked spots on the leaves. These spots later turn brown. Anthracnose spots also develop on cucurbit fruits, producing circular sunken areas of watery or soft-rotted tissue. Anthracnose can be controlled by spraying with Captan, or a copper based fungicide.

During warm wet weather, cucumbers may be attacked by downy mildew. Infected plants will develop yellow, irregular spots on the upper surface of leaves. A downy or cottony growth will develop on the lower leaf surface. Leaves turn yellow, then turn brown and die rapidly. The disease process begins in the older parts of the plants and progresses to the younger parts. Downy mildew can be controlled by spraying with the same fungicides that are used for anthracnose.

Blossom blight is another fungus disease that attacks cucumbers. This disease attacks the blossoms in humid weather. It causes the blossoms to wilt and fall off. The dying blossoms may be covered with a dense mold growth that is white at first, then turns brown or purple. Later, the fungus forms fruiting bodies that look like small pins stuck into the plant. The fungus also moves into young fruits, producing a soft wet rot at the blossom end.

Control blossom blight by growing cucumbers in well-drained soil. Annual crop rotation is a must in controlling blossom blight as well. Pick off and destroy any infected fruits or blossoms you find. A fungicidal spray used in the control of anthracnose and downy mildew might also help to reduce the damage from blossom blight but it will not completely control the disease without proper crop rotation.

Powdery mildew usually occurs later in the growing season, during dry weather. Powdery mildew is of little consequence when it occurs late in the season. However, if powdery mildew develops early in the season, it can be controlled by spraying with benomyl or similar fungicide.

Harvesting

Cucumber harvest is dependent upon their use. A safe rule of thumb is to harvest on the basis of size. Gherkins are harvested when quite small. Picklers can be harvested at any stage depending on their use. Slicing cucumbers should be picked when long and slender.

Slicing cucumbers should not be allowed to grow too large or reach the yellowish stage as they then are over-ripe and become bitter. Pickling cucumbers are sometimes allowed to grow larger and become yellow and are used in pickles and relishes that call specifically for yellow pickling cucumbers.

Harvest cucumbers by cutting the stem 1/4 inch above the fruit. Don't disturb the vines any more than necessary while harvesting the crop, as the vines will continue to produce new fruits. Frequent picking of cucumbers is essential as the fruits grow and reach optimum quality. As in other cucurbits, any delay in harvest can result in reduced yield and quality of fruit. Cucumber fruits grow quickly so the vines should be checked daily.

Eggplant

Eggplant is a member of the nightshade family, making it a close relative of the tomato, pepper and potato. Eggplant is a very tender annual that requires a long, warm season for successful production. The plants are killed by light frost and are injured by long periods of chilly, weather. Eggplant should not be set out until all danger of frost has passed.

Eggplant fruits come in many different colors, shapes, and sizes. The fruits can be used as a vegetable, a meat substitute, or in pickles.

Planting guidelines

Because the eggplant requires such a long growing season to achieve optimal production, the plants need to be started indoors at least 8 weeks before the last expected spring frost. You can purchase already started eggplant from your local nursery, although many nursery-started plants are root-bound and will become poor producers as they grow. It is best to start your own plants if possible.

Place 2 seeds per peat pot, cover lightly, and water well. Place in a sunny window to grow. Remove from window each evening to prevent chilling of seedlings. When seedlings are an inch tall, thin to one plant per pot.

Set transplants out in the spring after all danger of frost has passed. Place plants about 2 feet apart in rows spaced approximately 3 feet apart.

Maintenance

The use of black plastic mulch can increase the yield of eggplant by helping to warm the soil. It also helps conserve moisture and control weeds. Water eggplant during dry spells.

Insect pests and diseases

Insect pests of eggplant include flea beetles, Colorado potato beetle, tomato horn worm, aphids, and spider mites. Flea beetles are tiny and black. They eat numerous small holes in leaves, often peppering the leaf surface with holes, and can be particularly injurious to small plants.

Tomato horn worms are large, thick, green caterpillars that have ugly pointed horns on their back ends.

Colorado potato beetle adults and larvae both feed on eggplant leaves and can completely defoliate small plants if not controlled.

Hand pick the yellow and black striped Colorado potato beetles and large green tomato horn worms when possible. If you can pick the pumpkin-colored egg clusters of the Colorado potato beetle off the leaves and destroy them, you will go far in controlling this pest. Larval Colorado potato beetles are brick-red, and humpbacked. Hand pick them whenever you see them. Carbaryl, rotenone, and pyrethrins will also ensure some measure of control for all of these pests. Some *Bacillus thuringiensis* (B.t.) insecticides will control smaller larvae and are very safe to use.

Four-year rotations with non-related crops and using plants grown from disease-free seeds will help control most eggplant diseases.

A particularly damaging disease in eggplant is Verticillium wilt. This disease causes stunting in plants and a yellowing along leaf veins. The leaves also exhibit severe wilting, and finally death. Pull and destroy any plants infected with this disease.

Avoid planting eggplant anywhere tomatoes, potatoes, peppers, okra, raspberries, or strawberries have been grown for at least 5 years.

Harvest

The fruits of the eggplant are edible from the time they are one-third grown until ripe. They remain in an edible condition for several weeks after they become colored and fully grown. Skin should be shiny and the seeds inside should not be brown or hard.

The harvest will continue over an extended period if the fruit are removed when they are well-colored and of adequate size.

The fruits are usually cut from the plants since the stems are hard and woody. The large calyx, or cap, and a short piece of stem should be left on the fruit. Mature plants of most cultivars have sharp spines on them, so care is necessary when cultivating or harvesting to prevent personal injury.

Store eggplant in refrigerator. Use within a few days of harvest.

Garlic

Garlic is a versatile vegetable that is most often grown for use as an herb or spice.

Garlic grows equally well in both cool and warm climates. Garlic is most often planted in the fall throughout its growing range. In warmer climates it will grow throughout the fall and winter months and be ready to harvest in the late spring. In cooler climates garlic is planted in the fall just before the soil freezes. The garlic cloves remain dormant in the ground throughout the winter sprout up early in the spring. Garlic can also be planted in the spring for a fall harvest.

There are several different varieties of garlic available. There are two main types are distinguished by the names soft neck and hardneck garlic. There are several different cultivars within each type. Hardneck garlic is known as a top-setting garlic because of the woody flower stalk and the cluster of bulbils it produces after flowering.

Most hardneck types tend to produce large underground bulbs made up of a few large cloves. This type of garlic yields best when planted in the fall.

Yields can be increased by removing flower heads before the bulbils form. Save these tender young flower stems for use in cooking similar to the way young green onions are used. If left to grow, the bulbils, which are about the size of a popcorn kernel, can be eaten or they can be planted for another crop. If bulbils are used for propagation, it will take 2 to 3 years to produce a full-sized bulb. These bulbils can also be planted for garlic greens.

Softneck garlic does not form the woody stalks of hardneck garlic, but rather has long, flexible leaves that are often braided when the garlic is harvested and dried.

Bulbs of softneck garlic usually have more individual cloves and boast a higher yield than hardneck types. Softneck types also are generally better adapted to a wider range of climates. It is the softneck types that are most often planted in the spring.

Elephant garlic (*allium ampeloprasum*) is not a true garlic and is actually more closely related to the leek than to ordinary garlic. It is included in this category only because the average homesteader regards it as a form of garlic. Elephant garlic bulbs tend to be very large and can sometimes weigh over a pound. A single clove of elephant garlic can be as large as a whole bulb of standard garlic.

Elephant garlic is much less intense in garlic flavor and is sweeter. Elephant garlic has been described as "garlic for people who don't like garlic."

Planting guidelines

Plant garlic in rich, loose, well drained soil. Plant garlic bulbs directly in the garden soil in the same manner as onions. Don't crowd your garlic plants, but give them plenty of room to grow. Divide the bulb into individual cloves and plant each clove in the prepared bed spacing the cloves about 2 inches deep and 6 inches apart. Space rows at least 2 feet apart.

Maintenance

Water garlic only during very dry spells. In the driest climates, irrigation may be required, however be sure not to over water. Over-watering your garlic will only encourage disease processes. For most homesteaders, covering the garlic bed with a straw mulch will help the soil retain enough moisture for greatest yield without extra watering.

Remember to pick off the flower heads of hardneck garlic varieties for the highest yield unless you plan on using the bulbils for propagation or consumption.

Insect pests and diseases

Some of the insect pests that attack garlic are bulb mites, onion maggots, wireworms, pea leafminer, thrips, and wheat curl mites.

Onion maggots and wireworms can be prevented by good crop rotation. Never plant garlic where onions have been grown in the past year. Avoid planting garlic where sweet potatoes, alfalfa, oats, or wheat has been grown for 5 years. The typical symptom of onion maggots is the premature dying of the leaf tips. Control involves sanitation since insecticides are not effective against this pest. Onion maggot is usually only a problem following wet, cold periods on soils rich in organic matter, or soils that hold in moisture for long periods.

Thrips are very difficult insects to see. Adults are tiny and they usually feed deep in the neck of the garlic leaves where they are protected from natural enemies and pesticides.

Thrips abrade the outer layer of the leaves with their rasp-like mouthparts.

Leaves soon develop silvery blotches, and streaks. Hot, dry weather is favorable for thrips. Good garden sanitation can limit populations of thrips on your garlic. Destroy any plant residue after harvesting bulbs to prevent insects from over wintering in the garden. Spray when injury is first noticed. Rotenone and pyrethrum are usually effective against these pests. Generally however, good sanitation and crop rotation will be the best defense against these insects.

Diseases of garlic includes; bacterial soft rots, basal rot, black mold, blue mold rot, botrytis leafspot, botrytis bulb rot, downy mildew, pink root, purple blotch and stemphylium leaf blight, rust, sour skin, and white rot.

Most of these rots and molds can be controlled by good sanitation and providing good air circulation around each plant.

Keep weeds at a minimum in your garlic patch to improve air circulation around plants. Pull and destroy any infected garlic plants. Never add diseased garlic to your compost pile.

Disease resistant garlic cultivars can be purchased.

Harvesting garlic

Garlic is ready for harvest when the

bottom two or three leaves have turned yellow. Harvest garlic similar to the way you would onions. Using a small trowel or fork, gently loosen the soil around the bulb and ease the garlic out of the ground.

Regardless of variety, garlic needs time to cure and is not ready to eat until it has gone through 2 weeks of drying, called curing after harvesting. Brush the dirt off the plants and bulbs and lay them out on a screen, making sure that air can circulate around each plant. Cure the plants in a dry area in full or partial sun. Curing is complete when the skins are dry and the necks are tight. After the garlic is fully cured, you can cut the bulbs off from the rest of the plant. Store cured bulbs in a cool, dry environment.

You can also braid soft-neck garlic in the same way that you French braid long hair. Store the garlic braids in a cool dry place.

Cured garlic will last for 6 to 8 months in a cool, dry place.

Elephant garlic is more perishable than ordinary garlic so it doesn't keep as long.

Jerusalem Artichokes

Also known as Sun chokes, Jerusalem artichokes can be grown in most areas, but produces best in more northern climates. The plants are quite tall, often 6 to 12 feet tall, producing flowers that resemble their cousins, the sunflowers. The edible portion of the plant is the root, or tuber.

Fresh Jerusalem artichoke tubers taste similar to water chestnuts. The tubers can be eaten fresh, cooked like potatoes, or pickled.

Planting guidelines

One thing to remember when planning for your Jerusalem artichoke patch is that they spread rampantly, similar to horseradish. Keep your Jerusalem artichokes confined to a patch of their own to prevent them from choking out other desirable plants.

Jerusalem artichokes are propagated in a similar fashion to potatoes. Cut large tubers into chunks, or leave smaller tubers whole. Plant the tubers early in the spring about 6 inches deep, and about 1 foot apart in rows that are spaced 3 to 4 feet apart. Gently firm soil over tubers and water in well.

Maintenance

Try to keep weeds and grass to a minimum in the Jerusalem artichoke patch until the plants reach about 2 feet tall. After that the plants will be able to outgrow the weeds on their own. You won't need to cultivate between rows because cultivating disturbs the clusters of growing tubers. You can suppress weed growth by putting down a layer of straw mulch between rows. This will also help to conserve soil moisture and prevent soil compaction. Jerusalem artichokes are an easy maintenance crop.

Insect pests and diseases

Although Jerusalem artichokes suffer from most of the same pest and disease organisms that sunflowers do, Jerusalem artichokes have very few insect pests or diseases that affect them to any permanent detriment.

Unlike sunflowers, Jerusalem artichokes are perennials and will re-grow from the same tuber.

Sunflower maggot can sometimes infect your Jerusalem artichokes, as can various aphids, and beetles. Most insects do not seriously damage the plant to the point of plant death. Hand pick what insects you can, and dust with rotenone or carbaryl if insects become too much of a problem.

Most diseases of Jerusalem artichokes such as mildews and rots can be controlled by careful selection of the planting bed. Jerusalem artichokes should not be planted in low or wet spots. Good drainage is important to the health of these plants.

Good sanitation practices also aid in the control of disease. Remove and destroy any plants with symptoms of rot or mildew, such as yellowing, stunting, or cottony masses growing on the leaf surfaces.

Many homesteaders mow down their Jerusalem artichoke patch every fall. Rake up and burn the debris.

Harvesting Jerusalem Artichokes

The Jerusalem artichoke tuber should be left in the ground until fall, and is best after a light frost. It is dug in the same fashion as a potato.

The tubers are uneven and knobby with much dirt clinging to them. Wash them well, using a fine vegetable brush if needed.

Jerusalem artichokes can eaten raw or cooked, as desired.

Kohlrabi

Kohlrabi is a member of the cabbage family and like other members of that family, it prefers cool weather. Kohlrabi grows best and develops the best flavor when grown in cool weather, so in southern climates it will be best to plant it as a fall crop. The edible bulb of kohlrabi is actually part of the stem that sits just above soil level. This stem bulb swells out into this tender, sweet tasting vegetable which can be eaten either raw or cooked. Green, white, and purple skinned varieties all exist, however, they all are a creamy white inside.

Kohlrabi needs rich, well drained soil and full sun. Water frequently but do not allow the plants to stand in water.

Planting guidelines

Kohlrabi is an ideal plant for interpolating amongst other crops.

For a late spring or early summer crop, sow kohlrabi seeds outdoors as soon as the soil can be worked in the spring. You should plan to harvest your spring crop before hot summer weather sets in.

If you have a shorter growing season you can start seedlings indoors and transplant them a few weeks before the last expected spring frost date. For a fall crop, directly sow seeds into the garden. Plan so that you harvest a week or two after the first fall frost is expected in your area. When transplanting seedlings into the garden, space the plants about 6 inches apart in rows that are about 2 feet apart.

Direct sow kohlrabi seeds thinly 1/4 to 1/2 inch deep, in rows about 2 feet apart. Thin seedlings when they are an inch tall to stand six inches apart.

Maintenance

Kohlrabi likes rich soil, so side dress your plants with composted manure when the plants are about 6 inches tall. Cultivate to keep weeds down. Provide plenty of moisture but do not allow plants to stand in water. Kohlrabi needs well drained soil.

Insect pests and diseases

There are few insect pests that bother kohlrabi, partly because it is a cool weather plant and most insects need warmer weather. Kohlrabi is also a fast growing plant that is usually harvested before the usual cabbage family pests become a problem.

If cabbage loopers or worms become a problem they can be handpicked or the plants can be treated with carbaryl or rotenone or an insecticidal soap.

There are very few diseases that affect kohlrabi. Because of its fast growth rate and love of cool weather kohlrabi is generally considered disease free.

Harvesting

Harvest Kohlrabi when the bulb gets to range from three to five inches. Harvest the larger bulbs and leave the smaller ones to continue to develop and grow. To harvest just cut the bulb off at the base of the plant.

The leaves of the kohlrabi plant are also edible. They can be added to salads, sandwiches, or they can be boiled like other greens. The bulb can be sliced or cubed and boiled or fried lightly in butter.

Store harvested kohlrabi in the refrigerator for up to a week.

Leeks

See onions.

Lettuce

What would a green salad be without lettuce? You can grow a whole variety of lettuces to make your salads and sandwiches so much more appealing and healthful.

There are four main types of lettuce. Loose-leaf, cos, or romaine, butterhead, or bib, and crisphead. Looseleaf lettuce is probably the easiest type to grow, however none of them are difficult to grow. Lettuce varieties come in all sizes, shapes, and colors.

Lettuce prefers cool temperatures so it must be planted well before the last expected spring frost. Hot weather leads to bolting, or flowering. When lettuce bolts the leaves become bitter and unpalatable.

Planting guidelines

Leaf lettuce can be sown directly in the garden about 2 weeks before the last expected frost in the spring. It can be sown in rows or broadcast, or scattered, in the prepared lettuce bed.

All lettuce varieties need full sun and rich, well cultivated soil.

If you're sowing seed in rows, make a shallow furrow about ¼ inch deep. Sow the seed in the furrow, and then lightly cover the furrow, firming the soil gently over the seed. Lettuce can take a bit of crowding, so the seed can be spaced randomly in the furrow fairly close together. When the seedlings reach a couple of inches in height simply thin the seedlings to stand about 6 inches apart. Rows should be spaced about 2 feet apart. Be sure to use the thinnings for the salad bowl.

If you're broadcasting seed, simply scatter them atop the prepared garden soil, and then cover with about ¼ inch of fine soil or compost. After the seed germinates, thin the seedlings to stand about 6 inches apart. Again, be sure to use the thinning for salads, etc.

You can start any lettuce indoors in peat pots 4 to 6 weeks before the last expected spring frost. The only special requirement to starting lettuce seeds is that they need a large amount of light during the daytime hours. If a cold frame is available it would be best to put the pots in the cold frame during the day and bring them back indoors at night if the night if the weather is very cold. Sow lettuce seeds in peat pots at the rate of 2 seeds per pot. Using peat pots will prevent damaging the tap root when transplanting to the garden. Water the pots well and set them in a sunny window to germinate. Keep the pots evenly moist. Do not let them dry out completely. Keep the soil moist, but not soggy. When the seedlings get the first set of true leaves, thin them to one seedling per pot.

When the weather warms sufficiently, immediately after the last dangerous spring frost, set the seedlings out in the garden at the rate of one pot per every 6 inches for leaf lettuce or bib, or every 8 inches for head lettuce.

In warmer climates a second or fall crop may be planted.

Maintenance

Side dressing with a nitrogen rich fertilizer, composted manure, or liquid kelp will improve yield dramatically.

Daily harvesting of the outer leaves

will keep them from getting over-mature. Overly mature lettuce leaves develop a bitter taste.

Lettuce doesn't like the heat or drought of summer. A good layer of straw mulch will help keep the soil moist and cool for lettuce roots. Don't let the soil dry out, rather keep it as evenly moist as possible without letting it become soggy. Soggy soil leads to rot.

Frequent cultivation will keep the soil loose and hold down the weed growth. Just use caution as lettuce roots grow close to the soil surface and can be easily injured. Rake away mulch, cultivate around the soil, and replace mulch layer. Or you can cultivate the mulch layer into the soil if it is beginning to compost, and lay a fresh layer atop the newly cultivated soil.

Insect pests and diseases

Insect pests include aphids, leafhoppers, and flea beetles.

The best treatment for controlling aphids and leafhoppers in lettuce is to minimize weed growth in the vicinity of your lettuce crop. Introducing ladybugs to your garden may add a measure of control because ladybugs feed on aphids, however, don't expect all the ladybugs released to stay in your garden.

If flea beetles are a particular problem in your lettuce, try spraying the leaves with an insecticidal soap solution.

Mosaic virus and downy mildew are two diseases that occasionally attack lettuce.

Mosaic virus is spread by aphids, so controlling aphids will control this disease. Keeping down the weeds around your lettuce patch that attract aphids is one good way to keep mosaic virus at bay. Planting disease resistant varieties is also desirable.

Cool, moist conditions are necessary for downy mildew development. Excess moisture on the leaf surface is essential for spore germination and infection. Symptoms include the development of pale yellow regions on the upper side of older leaves with corresponding white fluffy growth, the spores of downy mildew, developing on the lower leaf surfaces. It is much easier to prevent downy mildew than it is to treat it.

Again, minimizing weed growth, and keeping the soil from getting too wet will help prevent or control these diseases. Pick out and destroy any plants that are diseased. Planting disease resistant cultivars is the best measure of controlling these diseases.

Harvesting

Keep lettuce picked daily to maintain fresh, mild flavored leaves. Pick leaf lettuce anytime from very young to nearly full grown. Discard older, outer leaves, and any leaves with insect damage.

Pick head and bib lettuces as soon as heads form nicely. It should be slightly firm, but have a little give when the head is squeezed.

Rinse leaves in cold water before storing. Lettuce can be stored in the refrigerator for several days without losing quality.

Melons; Crenshaw, muskmelon, honeydew, and watermelon

The main three types of non-watermelon type melons are grown in homestead gardens in the United States are Crenshaw, honeydew, and cantaloupe. There are many more types of melons grown around the world. These three and all their relatives love heat and are slow growing and slow maturing. Melons can be grown in most climate however, as there are now several short-season varieties to choose from for those who live in colder climates.

One note: Almost all the cantaloupes grown in the United States are really muskmelons. True cantaloupes are smooth-skinned French cultivars that are rarely seen outside of specialty shops here. Our rich orange-fleshed, fragrant and flavorful melons with the ribbed and netted rind are indeed muskmelons.

There are several different types of watermelon available to the homestead gardener. There are yellow fleshed and red fleshed varieties as well as seedless and seeded cultivars.

Planting guidelines

When choosing the site for your melon patch, choose a site that gets full sun and good air circulation. Melons simply don't grow or produce well at all if they're cold. All melons like sandy, well-drained soil. Prepare the soil with plenty of well-composted manure before planting. Applying about 4 ounces of household borax per 1000 square feet of melon bed will yield more flavorful melons.

Because melons require such a long growing season, it pays to get a jump start on the season by purchasing started melon seedlings from a nursery or greenhouse, or by seeding your own melons in peat pots indoors about 4 weeks before the last expected spring frost date. Sow 2 seeds per pot approximately ½ inch deep. Water well, and put in a sunny window to germinate. Water regularly. Never allow the pots to dry out.

Ideal temperatures for melon seed germination should be above 55 degrees.

After the seedling sprout you will need to keep them in strong sunshine to prevent the seedlings from becoming spindly.

About a week before transplanting into the garden, harden off the plants by setting them outdoors each day for a few hours, gradually increasing the length of time the plants remain outside. Bring them indoors each evening.

Melons can be planted in the garden in rows or in hills or mounds. After all danger of frost is passed, plant each pot in the prepared garden bed. Bury the peat pot completely beneath the surface of the soil, firming the soil around the plant stem. If plants are a little leggy the soil can be loosely heaped around the stem of the plant to provide some extra support. If using the mound or hill method, space 3 plants about 1 foot apart in hills that are spaced 5 feet apart. If using the row system, space plants about 12 inches apart in rows that are about 5 feet apart.

Maintenance

Lay a 6 inch thick layer of straw mulch down around your melon plants to maintain soil moisture and to discourage weed growth. Do not cultivate around melon plants as they have shallow roots that sprawl for some distance from the main plant. Hand pull weeds rather than using a hoe. A good mulch will go far in helping prevent weeds and necessitating tedious hand weeding.

If smaller melons begin to form along the vines after midsummer, pick them off the vines as they will not have a chance to grow and mature fully before the colder weather sets in, and they will only pull nourishment away from other, larger developing fruits. Try not to feel guilty when thinning fruit for it is a necessary chore to achieve proper fruit growth and flavor.

Insect pests and diseases

Cucumber beetles, squash bugs, squash vine borers, mites, and aphids are the most common insect pests affecting melons.

Cucumber beetles are oblong beetles that are yellowish-green in color, with three vertical black stripes down their backs. Rotenone gives good control for cucumber beetles. Cucumber beetles spread bacterial wilt and need to be controlled.

Adult squash bugs are rather large, a little over ½ inch long, winged, brownish black, and are sometimes mottled with gray or light brown, and flat-backed. Eggs are yellowish-brown to brick red laid in groups or clusters.

Squash bugs tend to give off a disagreeable odor when crushed. Both nymphs and adults suck sap from the leaves and stems of cucurbits, apparently at the same time injecting a toxic substance into the plant causing a wilting known as Anasa wilt, named for the insects scientific name, Anasa Nistis. After wilting, vines and leaves turn black and crisp, and become brittle. Small plants are killed entirely, while larger plants will have one or several runners affected.

Squash bugs are often found in large populations, congregated in dense clusters on vines and unripe fruits. Sometimes no fruits are formed.

Bacterial wilt is a bacteria that invades the vascular, or water conducting, tissues of cucurbits. It causes a rapid wilting of the plant. Cucumbers and melons are more often attacked than are squash and pumpkins. These are attacked only occasionally. Progressive wilting occurs, beginning with a single leaf and spreading to including the entire plant. The bacterial organism produces a sticky substance that blocks the water flow through the vascular tissues, producing the typical wilt symptom. Plant wilting and stringing of sap when affected stems are cut are diagnostic for bacterial wilt. The bacterial wilt organism is carried from plant to plant by the striped cucumber beetle.

Prevent bacterial wilt by controlling the cucumber beetle. Begin beetle control early, as cucumber beetles may attack as soon as seedlings emerge from the soil. Use carbaryl, or diazinon for control. Frequent light applications of pesticides are best, beginning when the plants are young to achieve beetle control as well as to avoid plant damage from the insecticide. Promptly pull and destroy any diseased plants.

Anthracnose is a severe fungal disease that spreads rapidly in warm, wet weather. The first symptoms usually appear on older melon plant leaves as small yellowish circular spots. The dead tissues turn brown in most melons, although in watermelons the dead tissues turn black. The disease spreads from older leaves to younger leaves. In warm, wet weather all the leaves may be attacked in rapid succession, giving the planting a "burned-out" appearance. Stems are also attacked, and light brown to black streaks develop. Circular, sunken, water-soaked spots develop on the fruits. These spots turn dark green to brown. In wet weather a pinkish ooze is produced within the spots. These pinkish spots are the fungus spores, which function like seeds. The spores are spread from plant to plant by running water, including splashing rain and by individuals working in the field when the vines are wet.

The anthracnose fungus overwinters on seed and on diseased crop refuse. Anthracnose control is difficult once the disease becomes widespread and serious. Prevention is the best control, including use of good quality seed, good sanitation practices, and appropriate crop rotation.

Powdery mildew is a common fungus disease of all melons. Powdery white spots appear on the upper surfaces of older leaves, usually beginning around mid-season or later. During hot, humid weather the disease can progress rapidly, with the upper surfaces of all leaves developing a white powdery appearance. Severe powdery mildew outbreaks cause the leaves to turn yellow and wither. Muskmelon, honeydew and Crenshaw fruits are not usually infected, but watermelon fruits are occasionally infected, with fruits becoming distorted or sun-burned due to loss of shade from leaf death.

Planting with ample air space between vines will help prevent powdery mildew from taking hold of your plants. Pulling and destroying advanced-diseased plants, and practicing good garden sanitation will also help prevent this disease.

There are several fungicides that can be purchased to help combat powdery mildew. Always read the label to be sure the fungicide is affective against this disease.

Mosaic viruses produce a patchwork or mosaic pattern of light and dark designs on the leaves and fruits of melons and other cucurbits. Leaves become small and puckered, and the plants are usually severely stunted. Fruits develop knobs or warts on them, and often the fruits are misshapen. Cucumber mosaic is very common. In addition to the mosaic pattern the edges of the leaves turn down, and the knobs on the fruits are light yellow. The cucumber mosaic virus is transmitted from plant to plant by aphids. Cucumber mosaic is readily spread from plant to plant on the hands of homesteaders working in the melon patch as well as by aphids.

Squash mosaic, which is caused by the squash mosaic virus, is transmitted from plant to plant by cucumber beetles, another reason to control this insect pest. The virus infects squash, cucumber, melon, and occasionally watermelon. The virus is also sometimes seed-borne.

Mosaic diseases are managed by using good quality seed and by controlling aphids and cucumber beetles throughout the season. Carbaryl is a good, low toxicity insecticide to use to control these beetles and aphids. Begin insect control as soon as seedlings emerge from the soil. Do not plant melons bordering woods, shrubby areas, or weedy areas. Controlling all weeds, especially perennial weeds, around your melon plants will also help deter aphids. Pull and destroy diseased plants as soon as mosaic virus appears. Good sanitation will help to reduce virus spread. After handling any diseased plants, wash your hands well with detergent and water to reduce the risk of spreading the disease further.

Although late season vine collapse is not truly a disease, it is included in this section because of it appears to mimic several different diseases. Muskmelons that mature late in the season are affected in some years by late season vine collapse. The plants may suddenly wilt and die before the fruits are fully mature.

Late season vine collapse usually occurs when sunny days follow cool weather. If soil temperatures drop to 50 F or below, the roots become inactive and cannot supply moisture to the plants. Plants with a good crop of fruit are easily moisture-stressed and may collapse rapidly if sunny weather follows cool weather. Losses can be minimized by not planting to late for a safe melon harvest. Remember, melons prefer warm temperatures.

Harvesting

Muskmelons should be harvested at the stage where you can press with your thumb at the point where the melon joins the stem, and the melon slips off the vine.

Crenshaw melons are ripe when they give off a sweet fruity odor.

The rind of casaba and honeydew melons turn yellowish when they are ripe.

A good way to determine if your watermelons are ripe is to thump the belly of the melon. If it gives off a hollow sound it should be ripe for picking. Also, a yellowing of the skin on the side of the watermelon that rests on the ground is another good indicator that the watermelon is fully ripe.

Melons do not have to be stored in the refrigerator, however, storing them in a colder place slows down the continued ripening of the fruits.

Okra

Originally a native of Africa, okra is a heat loving plant that grows best where growing seasons are long, however, it can be grown anywhere sweet corn will grow. By starting seeds indoors even northern homesteaders can enjoy okra.

Okra grows up to six feet tall and needs a lot of room to grow. It can be grown to form a hedge at the border of your garden. Okra grows rapidly, producing large, pale yellow flowers. The slender, pointed seedpods soon develop and are ready to harvest within 60 days of planting the seeds.

Planting guidelines

Choose a site that gets full sun, and where the mature plants won't cast shade on shorter crops planted nearby. In warmer climates sow seeds directly outdoors when the soil temperature has reached 65 degrees F. If the soil is any cooler than 60 degrees F. the majority of seed will simply rot in the ground. Soaking the seed in tepid water for a few hours before planting will enhance germination. Space seeds about 6 inches apart in rows that are about 3 feet apart. Thin seedlings to stand about 18 inches apart when seedlings are a few inches tall.

In more northern climates okra can be started indoors 6 to 8 weeks before the last expected spring frost. However, bear in mind that okra seedlings do not transplant well. When starting plants indoors, sow okra seeds in individual peat pots. Plant 2 seeds in each pot. After the seeds germinate, thin to one plant per pot. Transplant outdoors in the spring after all danger of frost has passed and when the weather is sufficiently warm.

Maintenance

Okra tolerates dry spells well. A good soaking once or twice a week, especially during dry spells, allowing the ground time to dry out between waterings, is all that is required.

Avoid the heavy use of any high-nitrogen fertilizers during flowering and pod formation or the pod yield will be low.

Insect pest and diseases

Aphids often attack okra plants. Rotenone and carbaryl are both effective against aphids, as is insecticidal soap. Often when aphids are present in the garden, ants will be present as well. Ants cultivate aphid colonies for the honeydew that aphids secrete from their bodies. If you have a large proportion of ants on your vegetable plants you may want to look around for signs of aphids.

Flea beetles, Japanese beetles, blister beetles, and cucumber beetles also attack okra. Hand pick insects whenever possible. The use of pyrethrins, carbaryl and rotenone will help to control these pests as well. Japanese beetle traps hung up outside the garden will reduce beetle populations inside the garden.

Corn earworm and cabbage worms both will attack okra. Again, hand picking and the use of insecticidal soaps and insecticides will control these insects.

Okra is generally considered to be disease resistant.

Blossom blight can be a serious problem in persistent rainy periods.

Root knot nematodes produce knots, called galls, on okra roots, which stunt and weaken the plants. Root-knot nematodes are actually microscopic worms that live in the soil and can be transmitted onto seeds. To control these nematodes, rotate crops with legumes or other plants not related to tomatoes, peppers, eggplant or okra.

Harvesting Okra

Harvest okra when pods are 2 to 4 inches long. (This is usually 5 to 6 days after flowering.) Use a sharp knife or hand shears to harvest the pods. Handle them carefully as they bruise easily. Pods mature quickly so okra should be harvested daily to ensure a tender, ample harvest. Pods that are more than 5 inches in length become tough and stringy. While the larger pods are still edible, the quality is usually considered quite unacceptable. Pods that have become too large to use should be promptly picked and discarded. Pods that are allowed to mature on the plant will reduce plant production.

Some individuals may be sensitive to the small spines on the okra plants' leaves and stems and may develop a irritated rash. Sensitive individuals should wear gloves and a long-sleeved shirt when harvesting the pods.

Okra should be used within a couple of days of harvest for best quality. Okra can be frozen, canned, or pickled.

Onions, Leeks, and scallions

Onions can be grown on nearly all types of soil from sandy loams to heavy clays. With the heavy clay soils some soil additives may be necessary. The addition of organic matter such as manure or other composted or decayed material will lighten the soil and increase the water holding capacity.

Onions are cool-season vegetables that can be grown successfully nearly anywhere. They are usually one of the first crops planted in the spring. Onions may be grown from sets, transplants or seeds.

Different varieties of onions require

different day lengths to begin bulb formation. Onions start bulb formation when the day length is of the proper duration for that variety of onion.

Most common varieties fall into one of three classes, long-day , grown in more northern climes, intermediate, which can be grown nearly everywhere, and short-day, which is suitable for the southern climates. As a general rule, onion varieties that are grown in the South are not adaptable to the North and vice versa unless an intermediate type onion is planted. Late plantings of the onion varieties runs the risk of small bulbs or even the complete lack of bulb formation.

There are several different cultivars of onion, each of which has its own culinary and medicinal uses.

Sweet Spanish onions are a long day type of onion of which there are several varieties. This type of onion is large, usually 4 to 6 inches across, globe shaped, white or yellow, exceptionally mild, fine grained, and sweet. Sweet Spanish onions can be cooked or eaten raw. They can be used either as a green onion or raised to maturity. These onions are usually grown from seed or from started plants. The biggest drawback to the sweet Spanish variety is that they are poor winter keepers, and are more susceptible to soft rot while in the ground.

Yellow onions are round to flattened onions that mature to about 2 ½ to 3 inches in diameter. They have yellowish-brown skins with creamy white flesh and have mild to strong flavors, depending on the variety. They can be grown for use as both a green onion and for mature onions. Yellow onions are excellent winter keepers. They are grown primarily from sets.

Bermuda onions are primarily short day varieties that are very mild flavored, and somewhat flattened on the bottom. Bermuda onions do not store well for winter use. They are grown primarily from seed or from started plants.

White storage onions can quite large or very small. They are generally a long day onion although there are intermediate varieties available. They have white papery skins and a sharp, clean flavor that ranges from mild to strong, depending on the variety. The keeping quality of white onions is not as long as that of the yellow onion.

Pearl onions are short day varieties that can be grown in yellow, white, or red colors. They form tiny bulbs suitable for pickling, creaming, or boiling.

Yellow granex is a short day variety that is responsible for producing some of the sweetest yellow onions known. They are somewhat flattened onions with yellow skins and cream colored flesh. Yellow granex is an extremely sweet onion but it has poor keeping qualities.

Leeks are grown for their long white bulbs. These bulbs will turn green and tough if they are exposed to the sun, therefore, leeks are usually planted in furrows or trenches so that the bulbs can be kept covered with soil.

Planting guidelines

Onions should be planted early in the spring as soon as the soil can be worked. If onion sets are used, they should be spaced 2 inches apart in the row and later thinned to stand 4 inches apart. Use the thinning as green onions. Sets should be planted 1 to 2 inches deep in rows that are set about 2 feet apart.

Prepare trenches for seeding leeks in the fall by digging a furrow about 6 inches deep and about 9 inches wide. If you dig more than one trench, leave about 2 feet of space between trenches. Clear the trench bottom of any stones or other debris, then, work about 2 ½ pounds of well composted organic matter and manure into each 12 feet of trench.

Start leek seeds indoors about 10 weeks before last expected spring frost.

Sow 2 seeds in each individual peat pot about 1/8 inch deep. Water well, and put in a sunny place to germinate. When seedlings appear, thin to one plant per pot.

In the spring, after all danger of frost is past, set seedlings out in holes dug in the bottom of the prepared trench. Dig holes about 6 inches apart. Plant pots deeply so that only the upper leaves stick out of the soil. Water well.

Maintenance

Keep weeds down by frequent and shallow cultivation and hand pulling. Weeds interfere with onion growth by crowding and moisture use.

Water onions well once weekly in dry periods.

As leeks grow, continue to heap soil up around the growing plants in the trench to keep the long bulbs well covered.

Insect pests and diseases

The two main insect pests that attack the onion family are onion thrips and onion maggots. These insects rarely infest the homestead garden, however, it is possible, especially if near a larger onion, leek, or garlic growing operation.

Onion thrips are small, yellowish,

sucking insects that attack the leaves giving them a blanched appearance. The center leaves become curled and deformed and the outer leaves turn brown at the tips.

Onion maggots are the tiny larva of a small fly. The flies lay eggs on the plant near the root or in cracks in the soil. The small maggots, only about 1/3 inch long, kill the young plants and then burrow into the bulbs. Their tunneling makes it easy for decay organisms to enter the onion bulbs and cause them to rot.

There is little that can be done to treat onion plants that have been infested with onion thrips or onion maggots. Prevention is the best treatment. Using proper garden sanitation as well as appropriate crop rotation should go a long way toward keeping your onion patch free from these pests.

There are many diseases that attack the onion family.

Onion blast is a disease that affects the foliage of the plant. It is a fungus disease that occurs rapidly, quickly devastating the plant's foliage.

Onion neck rot and bacterial soft rot are two rot diseases that affect the bulbs and they can be a major impact on mature onions that are grown for storage.

Onion neck rot is a softening of the

scales which usually begins at the neck but can occasionally start from a wound on the bulb. There is a definite margin between the healthy and diseased tissue. If the bulb neck feels soft and spongy, then in all probability the bulb is infected with Neck Rot.

Proper handling of onions is the most important control measure. It is vital that the neck tissue be dried out promptly and thoroughly upon harvest, which will check the growth of the fungus permanently.

Bacterial soft rot usually starts at the neck of the bulb but, unlike Neck Rot, progresses further into the bulb layers. An offensive odor is given off by the rotting bulb.

Again, proper handling helps prevent the rot from occurring. The organism that causes Bacterial soft rot enters the onion through a wound in the flesh. Moist conditions encourage the disease growth. Onion maggots are usual causes of wounding in onion bulbs, allowing the bacteria to enter. The insect also carries the bacteria from plant to plant as it feeds.

Harvesting

Onions for use as green onions can be harvested as soon as they reach a desired size. Homestead gardeners should plant enough onions to have for use as green onions as well as plenty for storage. Remember when thinning small onion seedlings to save the thinning for use as green onions.

Onions that are grown to be stored should be harvested when most of the tops have broken over. Digging of onions and handling after harvest should be done with care because any wound to the bulbs gives ready access to decay organisms.

The harvested onions should be placed on the ground in rows. Try to keep the bulbs covered as much as possible by the onion tops to prevent sunscald. The onions are left in these rows until the tops become dry.

You can also place newly harvested onions in a warm, dry place, out of the sun, such as a garage to dry. The length of time required for the tops to dry depends on the weather and may be anywhere from 3 to 10 days.

After the tops are dried they can be cut off. Leave about 1 inch of top attached to the bulb. If the top is cut too close, decay organisms have easy access to the bulb. As the bulbs are topped, discard any onions that show even the slightest sign of decay, or those that have wounds or have thickened or soft necks. Storage onions must be cured as well as dried before they can be safely stored. The easiest method of curing onions for the homesteader is to place the onions in mesh bags and tie the sacks to the rafters of a building that has free movement of air. A garage or shed works well if the door is allowed to remain open for air circulation. The onions should be allowed to cure for 3 to 4 weeks in this manner.

Storage onions should be kept in a cool, dry place, an unheated attic or dry storage room would be ideal. Onions can handle temperatures as low as 32 degrees F without any ill affects.

If desired, the dried tops can be left on the onions and they can be braided and hung in a dry place such as an unheated attic.

Leeks can be harvested as desired from the garden. They can be eaten when quite small, or left to grow larger. When harvesting leeks dig up the plant with a spading fork, taking care not to injure plants.

Leeks can be stored in sandy soil in the root cellar, or they can be left in heavy mulch in the garden for later harvest. Remove any side shoots that form on the plants and transplant them for another crop.

Parsnips

The parsnip is a root vegetable that takes a full four months to mature. They look like big, white carrots and have similar growth habits. A frost actually enhances the robust sweetness of the vegetable, making it a favorite for northern gardeners.

If the soil is mulched well enough to prevent actual freezing of the soil, parsnips can be left in the ground and harvested throughout the winter months. If the ground does freeze, the parsnips can be harvested the following spring. In warmer climates, parsnips can be planted in the fall for use as a winter crop.

Planting guidelines

Parsnips need deeply tilled, rich, loose soil. Sow parsnip seeds directly into the garden after the danger of heavy frost has passed, usually about 2 weeks before the expected last frost date. Sow seeds ½ inch deep and ½ inch apart in rows that are about 2 feet apart. When seedlings are about 2 inches tall, thin the plants so that they are 2 to 4 inches apart.

Parsnip seeds germinate very slowly, so it may take as much as 3 weeks for the seed to sprout. Parsnip seed can be soaked in water overnight to help hasten germination. A light layer of mulch spread over the seeded parsnip bed will help to keep the soil cool and moist which will encourage optimum germination. Radish seed may be scattered in amongst the parsnip seed to mark the row during the long germination. Harvest the radishes as it reaches desired size, leaving that room for the growing parsnip seedlings.

Maintenance

Adding mulch to the row of growing plants throughout the season will help control weeds and conserve soil moisture. This mulch will add up over the growing season and will help insulate the plants over the winter to facilitate continued harvest in the winter months.

Insect pests and diseases

Parsnips are not often bothered by insect pests.

Occasionally aphids will attack plants and cause foliar damage from their feeding. Insecticidal soaps, carbaryl, and rotenone are all fairly effective against aphids. Insecticidal soaps will have to be used more often as its residual effect is limited. A natural approach to aphid infestation is the use of lady bugs, which can be purchased through many garden supply catalogs.

Swallowtail butterfly larva like to feed on all members of the carrot family, including parsnips. Pick off these yellow and black striped caterpillars whenever you see them. One caterpillar can defoliate a plant in only a day or two.

Another parsnip insect pest is the carrot rust fly. This insect injures the roots of parsnips. This slender fly is about ¼ inch in length and has a metallic blue-black body. It lays its eggs in the soil close to the plant in late spring, and the young maggots work their way downward along the root and begin feeding at the tip of the root. Older carrot maggot larvae tunnel into the lower third of the main root. The maggot burrows are rusty brown in color. The maggots pupate in the soil and there are two generations of the carrot rust fly each season. The first generation maggots feed from June to July and the second generation from late August into September.

Not a significant problem in all areas, carrot rust fly can be controlled by using diazinon at the time of sowing as a preventative measure in areas prone to the fly. A row cover of fine nylon mesh is also effective in preventing the fly from infesting your crop.

Proper crop rotation and companion planting will also diminish the threat of this insect. Do not plant parsnips near carrots or celery.

Parsnips are prone to several different diseases. Luckily, none of these diseases will cause much of a problem for the homestead gardener. They may become a problem if you are raising larger crops for market, or if you live in a particularly disease prone area. For this reason, these diseases are mentioned here.

Root and crown rot, caused by fungi, have symptoms that cause the plant to appear as wilted with a subsequent collapse of the plant. The roots can appear brown and water-soaked instead of white. A water-soaked lesion can often appear at the base of the stem. Control can be achieved by using a two-year crop rotation with non-susceptible plants, such as corn, to prevent the buildup of the organisms in your soil.

Sorehead is caused by a fungus and causes damage to the roots, crown, and leaves. The diseased parsnip roots show dark, irregular, and slightly sunken pits. These cankers are usually limited to the top of the parsnip root, but may extend toward the tip of the root in severe cases. The fungus can also infect the leaf and cause a minute spot which can enlarge to form an irregularly-shaped lesion. The spores of the fungus wash down and infect the roots, where the cankers are produced. Good drainage and rotations with other crops for periods of six years can provide some control. There are also resistant cultivars that can be chosen in areas prone to this fungus.

Leaf spots are caused by fungi and appear as small angular spots with yellowish-green to dark brown spots on the foliage. These spot can grow together into one large area and cause the entire leaf to drop off. Control can be achieved with the use of fungicide sprays applied as soon as symptoms are visible. More natural control can be obtained by good garden sanitation and immediate removal and destruction of infected plants.

Powdery mildew is also caused by a fungus, and it appears as white powdery patches on the upper and lower surface of the leaves. These patches eventually grow to cover the entire leaf surface, and can cause leaf browning. Do not work amongst parsnips when the weather is wet to prevent the spread of mildews and other diseases. Removal and destruction of any infected plants will help prevent large outbreaks of powdery mildew. Do not overcrowd plants to allow room for plants to dry out between waterings.

Parsnip blight is a disease caused by soil-borne bacteria that causes the root interior to turn brown. Control can be achieved by practicing good crop rotation using two-year plan.

Aster yellows is another disease that strikes parsnips. Parsnips that have aster yellows have an abnormal number of leaves. The leaves are yellow, twisted, and stunted. The roots remain slender and have an abundance of fine hairy roots. Parsnip yellows is caused by a phytoplasma which also causes lettuce yellows, carrot yellows, and aster yellows. Phytoplasmas are carried by leaf hoppers. Control of these insects will control the disease. Leaf hoppers can be controlled with insecticides such as carbaryl, rotenone or insecticidal soaps. Avoid planting asters or carrots near parsnips.

Root knot nematodes sometimes infect parsnips. Infected plants are stunted and sickly and have knots on their small, feeder roots. This disease is caused by nematodes, which are small worms that can persist in the soil for years. Good crop rotation with non-susceptible plants, such as corn, can reduce the number of nematodes living in the soil. Growing parsnips in a new area will also control the disease.

Harvesting

Dig the roots with a shovel, tilling spade or spading fork. Roots will average from 1 to 2 inches in diameter and up to

12 inches in length. Yields frequently exceed one pound per foot of row. Harvest only what you intend to use and leave the rest in the garden bed, covered with a thick layer of mulch for future harvest throughout the winter. Parsnips can also be harvested and placed in sand in the root cellar for easy access throughout the winter.

Peas

Peas are frost-hardy, cool-season vegetables that can be grown nearly everywhere. Peas are planted early in the season and mature quickly.

Peas can be classified as garden peas or English peas, and the edible pod peas such as snap peas and snow peas, or sugar peas.

Garden pea varieties can have either smooth or wrinkled seeds. The smooth-seeded varieties tend to have more starch and be less sweet than the wrinkled-seeded varieties. The wrinkled-seeded varieties are usually preferred for home use. The smooth-seeded types are used more often to produce peas for split-pea soup.

Snap peas have been developed from garden peas to have low-fiber pods that can be snapped and eaten fresh along with the immature peas inside. Snow peas are meant to be harvested as flat, tender pods before the peas inside develop at all.

Both garden peas and edible pod peas come in tall-vined and dwarf varieties. The dwarf varieties require no staking or support, while the tall-vining varieties require some support on which to climb. The climbing or tall-vining varieties usually provide larger yields than do the dwarf varieties.

Peas are extremely sensitive to heat and will stop growing or producing when temperatures regularly reach above 70 degrees F.

Planting guidelines

Plant peas early in the spring whenever the soil temperature is at least 45 degrees F, and when the soil is dry enough to cultivate without its sticking to garden tools.

Sow peas from 1 to 1–1/2 inches deep and about one inch apart in either single or double rows. Allow 18 to 24 inches between single or pairs of rows. Allow 8 to 10 inches between each set of double rows.

Maintenance

Peas should be mulched to help cool the soil and retain moisture. Mulching also helps keep down weeds as well as soil rots. For vining types you may need to occasionally train the growing vines onto the support system.

Insect pests and diseases

Peas are not bothered by many insects because they are generally grown in the cooler weather, before most insect pests become a problem.

Peas that are grown in poorly drained soils are more susceptible to fusarium wilt and root rots. Symptoms are the yellowing and wilting of the lower leaves, as well as stunted growth of the entire plant. Infection of older plants usually results in the plants producing only a few poorly filled pods. Fusarium wilt can be avoided by growing wilt-resistant varieties.

Harvesting

For English peas, when the pea pods are swollen and appear well rounded they are ready to harvest. Pick a few pods every day or two as harvest time approaches to determine when the peas are at the proper stage for eating. Peas are of the best quality when they are fully expanded but still young, or before they become hard and starchy. Peas should be picked immediately before cooking because their quality, like sweet corn, deteriorates rapidly. The pods on the lower portion of the plant usually mature first. Pick over the peas every other day for a week.

At the final harvest pull up the entire plant to make picking easier.

Snap peas should be harvested every 1 or 3 days to get peak quality. Sugar snaps are at their best when the pods first start to fill out, but before the seeds grow very large at all. At this point, the pods snap like green beans and the entire pod can be eaten. Some varieties have strings along the seams of the pod that must be removed before cooking, similar to string beans. Sugar snaps left on the vine too long begin to develop tough fibers in the pod walls. These pods should then be shelled and used in the same manner as garden peas, discarding the pods.

Vining types of both sugar snap and snow peas continue to grow taller and produce peas for as long as the plants stay in good health and the weather stays cool.

Snow peas are generally harvested before the individual peas have grown to the size of a BB and when the pods have reached their full length but are still quite flat. This stage is usually reached 5 to 7 days after blossoming. Snow peas must be picked regularly (at least every other day) to assure sweet, tender pods. Snow pea pods can be stir-fried, steamed or mixed with oriental vegetables or meat dishes.

Keeping the plants picked will encourage the plants to produce longer. Like snap peas, overlooked peas can be shelled and used as garden peas, and the tough pods discarded.

Pea pods can be stored in a plastic bag in the refrigerator for two weeks. Unlike fresh green peas, pea pods deteriorate only slightly in quality when stored.

Peppers

Peppers are slower growing, warm season members of the nightshade family, along with tomatoes and potatoes, as well as others.

Peppers can be found in hot and mild varieties and come in a rainbow of colors from red, green, orange and yellow to the newer chocolate browns.

The sweeter bell type peppers are eaten fresh in salads and garnishes and are used for, soups, stews, sauces, and for relishes and pickling.

Hot peppers are increasingly used in cooking ethnic foods as well as their use as a seasoning and in pickles.

Hot peppers need to be isolated from sweet peppers or all your sweet peppers will be hot!

Planting guidelines

Because of their slow growing nature

and love for warmth, peppers are best started from seeds indoors in late winter and then transplanted into the garden after the soil and air have warmed in the spring. The plants cannot tolerate frost and do not grow well in cold, or wet soil. Start seeds indoors in individual peat pots 8 weeks before last expected spring frost. Sow 2 seeds per pot, cover seeds with soil and water well. Set pots in a sunny window to germinate. When seedlings reach 1 inch tall, thin to the strongest plant per pot.

Set transplants out in the garden when the soil has warmed to at least 50 degrees. Set plants about 24 inches apart in the rows that are also set 24 inches apart.

Maintenance

Side dress the plants with well composted manure in early summer when the first crop of peppers are set on the plants. Keep a good layer of compost around plants as peppers need continuous moisture but should not stand in soaking soil. A good mulch layer will help the soil retain moisture and keep it evenly moist as well.

Insect pests and diseases

Aphids can be common pests of peppers and can spread mosaic virus to your plants. Control aphids by the use of carbaryl, rotenone, insecticidal soaps or pyrethrins. Keep weeds around your garden mowed to discourage these pests further. Lady bugs will also help keep aphid populations at a manageable level. Take care when using insecticides with other natural approaches such as importing lady bugs as the insecticides will kill the desirable insects as well as the European corn borers are caterpillars that may feed on peppers, particularly if corn and pepper are planted near one another. The larvae tunnel into the flesh of the pepper fruits. The larva is pale white or gray with black tubercles and is no more than 1 inch long when fully grown. Adults have a wingspread of an inch or so and are tan or brown. Control begins with proper tillage of plant residue at the end of the growing season. This reduces the population of borers from the previous year. Bt formulations are also used to control this pest. Consult the label for usage directions, and pre-harvest treatment intervals.

The parasitic wasp Trichogramma has been used as an organic method of control. This tiny wasp attacks the egg masses of various caterpillars, including the corn borer. Chemical insecticides will harm these wasps, so if you are using natural insect predation methods of control you will want to limit your chemical insecticide usage.

Flea beetles are small, black, very active beetles may feed on plants early in the season. They are usually more of a nuisance than a serious threat. Control is not usually needed.

Pepper maggots sometimes infest the fruits of peppers, causing them to decay. The eggs are deposited in late July and August in the wall of the fruit. The adult fly is light yellowish-brown in color, with three brown bands across each wing. Serious infestations are usually limited to fruit developing during the brief period when egg laying occurs, and are usually limited to commercial operations and gardeners adjacent to them.

Damping off is a disease that causes a rotting of the roots and the base of the stem so that the seedling literally falls over and dies. Damping-off is caused by a number of different fungi which live in the soil. The disease is encouraged by high soil moisture and cool temperatures. Sow seeds in pasteurized potting soil or in a soil-less potting mixture will start the seedlings off well. The use of fungicide-treated seed can also reduce damping-off. Be careful never to over-water peppers. Transplanting to evenly moist soil only after the weather has warmed sufficiently in the spring will also help prevent damping off from occurring. Once plants are beyond the seedling stage, they are no longer susceptible to damping-off.

Phytophthora blight is another disease that occurs in wet conditions. This blight is a two stage disease with the first stage appearing as a crown rot that is characterized by stunting, yellowing, and wilting. The second stage is characterized by blackened lesions which coalesce on the foliage and eventually kill the entire plant. To control this fungal disease plant in well drained soil and practice good sanitation and proper crop rotation. Do not include tobacco, tomatoes, eggplant, or potatoes or any of the squash family in the crop rotation. Fungicides are, for the most part, ineffective against this disease.

Pythium root rot is a fungal disease that resembles the crown rot phase of Phytophthora blight. Plants appear stunted and wilted with water-soaked tissue at the stem base. The disease can appear in wet, poorly-drained soils. Again, control measures include planting in well drained soil and practicing proper crop rotation, omitting any nightshade relatives from the rotation schedule.

Anthracnose, another fungal disease, causes dark, circular, sunken spots to appear on either green or ripened fruits. These spots may be in inch or more in diameter. The surface of the spots is covered with black dots that contain the spores of the fungus. After rain or heavy dew, pinkish masses of spores exude from these dots. Infected fruits may be completely rotted away and fall from the plants. They may also cling to the plants as withered "mummies." Diseased fruits serve as sources of infection for other fruits.

Anthracnose is a soil-born and seed-carried disease which also infects and survives on stems and leaves throughout the growing season. This disease often goes unnoticed until the fruits develop because symptoms on stems and leaves are usually very slight. It is important to be observant while working in your garden to watch for the earliest sign of disease. Good garden sanitation will help control this disease. Pull and destroy any plants showing symptoms of this disease. There are fungicides that are labeled to treat this problem. Always follow label directions and follow safety precautions. Again, maintaining even soil moisture and proper crop rotation will help prevent this disease from taking hold

Bacterial spot is a bacterial disease that appears on stems, leaves, and fruit.

Potatoes do particularly well in new garden soil. Potatoes require acid soil, so do not plant them in any area that has recently been treated with lime.

Planting guidelines

Potatoes can be purchased as sets, or precut potatoes, or you can prepare your own seed potatoes for planting. Do not try to use potatoes purchased from the grocery store to seed your potato crop. Store bought potatoes are often treated to prevent sprouting, and are not certified as disease free.

To prepare your own seed potatoes for planting, cut them into pieces that are about the size of a walnut. Be sure each piece has at least one eye. After cutting, spread the seed pieces out to cure in a bright, airy place until the cut surfaces harden and dry slightly.

As soon as the ground can be worked in the spring, plant the seed pieces in furrows that have been dug to a depth of 4 inches deep and 3 inches wide. Keep furrows about 3 feet apart. Set the potato pieces about 12 inches apart with the eyes facing upward. Cover the potatoes with 3 inches of soil.

Potatoes take about 3 weeks to sprout, but the plants grow fairly quickly once sprouted.

Maintenance

As the plants grow, use a hoe to mound the soil up around the base of the growing plants to give support to the vines as well as to keep the developing tubers well covered with soil. Potatoes that are exposed to sunlight turn green and develop a poisonous substance called solanine. Mulch with straw to help retain soil moisture and keep down weeds.

Insect pests and diseases

A potato plant's worst insect enemy is the Colorado potato beetle. Both the yellow and black striped adult and its red, humpbacked larva will defoliate and destroy potato plants in a very short time. Hand pick these pests and use carbaryl or pyrethrins to control them. Aphids, leafhoppers and flea beetles will also often attack potato plants. Insecticidal soaps and carbaryl, rotenone and pyrethrins can all be effective in the control of these insects.

You can prevent many potato diseases, such as potato scab, by using certified seed potatoes and keeping the soil acid.

Potato blight first appears as purplish blotches on the leaves, which then turn brown and rot. Blight diseases are caused by fungi and are the most damaging disease to affect potatoes. Blights are more common in wet, humid weather. To control this disease, keep potatoes weeded to allow for ample air circulation. Practice good sanitation and proper crop rotation. Burn all infected garden residue in the fall. Again, plant only certified seed potatoes.

Harvesting

You can start harvesting your potatoes around the time the plants blossom. Push aside the soil at the base of the plant and carefully pick off some of the new potatoes. These are highly valued tiny "new" potatoes that can be cooked whole, in their tender skins. Be sure to leave some potatoes to grow to maturity.

When the plant foliage begins to wither and die back, the potatoes will be full grown. Dig potatoes from the soil with a spading fork or a round nosed shovel. Potatoes can be left in the ground for a while after the foliage dies back, however, they should be harvested before the first heavy frost hits. Potatoes should be cured to harden the skins by setting them in a dark, dry place for a week or two. Sort through potatoes before storing them to remove any potatoes that were damaged during harvest.

Radishes

There are two types of standard radish. The popular red or white summer radish that is fast maturing and highly flavored, and the slower-maturing, sharper-flavored winter radish that has either a white or black skin. Within these two types there are hundreds of varieties. Radishes like cool weather and will thrive nearly everywhere. Radishes are among the easiest and fastest vegetables to grow.

Growing guidelines

Cultivate the soil to a depth of 6 inches and rake soil smooth. Like most root crops, radishes should be sown directly in the garden. They do not transplant well.

Sow seed 4 to 5 weeks before the last expected spring frost. Place seeds 1/2 inch deep and 1 inch apart, in rows or patches. Rows should be placed at least one foot apart for smaller varieties, two feet apart for larger types.

For continuous crop, make successive plantings, in the same or other beds, every 7 to 10 days until the average air temperature reaches 65 degrees F.

The less common winter radishes mature in 60 days or more, and they are planted in midsummer for fall harvest.

Garden space can be conserved by inter-planting radish seeds with other, slower-growing vegetables, such as carrots and parsnips.

Maintenance

Thin seedlings to two inches apart and spread a layer of mulch to suppress weeds and retain soil moisture. Overcrowded radishes will never develop into decent sized roots. Save thinning for use in the salad bowl. Cultivate and weed carefully to prevent damaging the roots.

Insect pests and diseases

Insects rarely attack the pungent radish. The main principal pest is a root maggot, which tunnels into the roots and is often found in the soil where cabbage crops have been grown in the past. Diazinon granules sprinkled on top of the rows after seeding will help control this pest.

Cabbage worms may also attack radish leaves. Bt, carbaryl, pyrethrins, or handpicking will control these insects.

Because they are so fast growing, maturing in as little as 20 days from seeding, radishes are not often attacked by disease.

Harvesting

Begin harvesting radishes as soon as they are big enough to use. Be sure to harvest before they become too large or woody. The will become bitter and tough if they are allowed get too big. A split or crack down the root signals over-maturity. All radishes should be harvested before flower stalks emerge.

To harvest, pull plants completely out of the ground, remove tops and discard, or save to use as a vegetable. Radish roots can be eaten fresh, in salads, or they can be boiled. Store harvested radishes in the refrigerator for up to 2 weeks.

Rhubarb

Rhubarb is generally classified as a vegetable, although most folks think of it as a fruit because of its use in pies, jellies or in a compote.

Rhubarb is a long-lived perennial that is planted in a permanent bed. Established rhubarb beds need a minimum of care and will produce

For many years. Rhubarb grows best in areas where winters are cold enough to freeze the ground to a depth of at least 2 inches.

Rhubarb grows in a variety of soils, however, it is a heavy feeder and does require copious amounts of organic matter and fertilizer.

Planting guidelines

Rhubarb is seldom started from seed. Usually it is propagated by planting root divisions known as crowns. Rhubarb crowns can be ordered from seed catalogs, or you can buy crowns and started plants from your local nursery or garden center.

Because Rhubarb is a perennial plant it will inhabit the same spot for many years, therefore, it should be given a secluded spot in or near the garden where it won't interfere with any other vegetables. If you don't want to plant your rhubarb in your vegetable garden, because of space restrictions, you can plant it along the border of your strawberry patch.

Prepare the spot by digging a hole for each crown. Dig holes 2 feet deep and 2 feet wide. Space holes 2 feet apart in all directions. Fill the bottom of each hole with a 6 inch layer of compost or manure. Mix the soil you have dug up with an equal amount of well composted manure or other organic material. Fill the hole to a depth of 1 foot with this soil mixture.

In early spring place one rhubarb crown in each hole so that the top, where the plant buds are located, sits about 3 inches below the soil surface. Firm the soil around the roots. Back fill each hole with the rest of the compost and soil mixture until level with the surrounding soil.

Maintenance

When the first growth appears, and every succeeding spring thereafter, spread about a half a pound of composted manure, or 10–10–10 fertilizer around the shoots, working it into the soil surface with a hand cultivator. Lay and maintain a mulch layer around and between the plants to retain soil moisture and discourage weeds. This mulch layer will also protect roots from the most severe winter freezing.

Rhubarb plants will send up flower stalks which will reduce stalk yield if allowed to flower. Cut off this seed stalk as soon as you notice them.

After a few years, your rhubarb patch might become over-crowded and the stalks noticeably thinner. This is the time to divide your roots. Dig up the plants in the spring when the new shoots are just emerging, or else early in the fall when the plants are no longer actively growing. Cut the roots into several parts, each of which should have one to three buds. Simply treat these sections like they were new crowns, planting them in the same manner. You can enlarge the existing rhubarb bed, or plant a second one. If you divide and replant roots in the fall, be sure to mulch heavily to help protect the young crowns.

Insect pests and diseases

Rhubarb is relatively free from insect pests and diseases.

The rhubarb curculio is a short, yellow beetle that sucks the juice from the plants. It thrives in weedy areas, so keeping the area weed-free and any grassy edges along the rhubarb patch mowed should go far in controlling this pest. Hand pick any beetles you see.

Harvesting

Do not harvest any rhubarb stalks from your plants the first year after planting. You can begin harvesting a few stalks the second year if the stalks reach close to 18 inches in height. Beginning the third year, harvest about half the stalks, leaving the thinnest ones to nourish the developing roots.

Harvest rhubarb stalks by grasping the stalk near the base and twisting it off. Never eat the tops as they are mildly poisonous.

Store rhubarb stalks in the refrigerator for no more than a week for best quality. Rhubarb can also be frozen for winter use.

Squash and pumpkins

Despite the differences in taste and appearance, the two main types of squash, summer and winter, are closely related and are grown in the same manner. They are cousins with the cucumbers, gourds and melons.

Summer squashes usually grow as large bushy plants whose fruits are harvested long before they reach full maturity. The rinds of most summer squashes are soft and edible.

Winter squashes grow as vines, requiring ample space to sprawl. The fruits of winter squashes are left on the vine to reach full maturity before being harvested. Their rinds are tough and inedible. Properly stored, winter squashes can be kept throughout the winter.

Pumpkins are actually a type of squash. Some pumpkins grow on vines, others on bush type plants. Like winter squashes, pumpkins are allowed to fully ripen before being harvested.

All squashes need rich, well drained, loamy soil.

If you grow a number of varieties of squash and you save the seeds, unless you separate varieties by several hundred feet, you'll probably find that next year's crop will produce some very strange and interesting mutations.

Planting guidelines

Squash is usually planted in mounds or hills, although it can be grown in rows as well. To prepare a hill, dig a hole 12 to 18 inches deep and about 2 feet in diameter. Fill the bottom of the hole with about 6 inches of well composted manure. Shovel the excavated soil back into the hole until it forms a mound that rises about 8 inches high. Space the hills 4 to 6 feet apart for bush varieties and 8 to 10 feet apart for vining plants such as winter squash and pumpkins.

Squash can be direct seeded in the garden, or started indoors about 4 weeks before the last expected spring frost date.

To direct seed, sow in rows or hills, planting seed one inch deep. Row spacing is dependent upon the variety you are planting, but averages one seed about every 6 inches.

In hills, plant four to five seeds per hill. Cover with soil and firm the soil lightly. Water the first day and if there is no rain, every two to three days until the seeds germinate.

Vining winter squash can be seeded between rows of peas, so that as the peas are harvested and the plants removed, the squash vines can grow to make double use of the same space.

Some winter squash, such as jumbo pink banana, hubbard, and pumpkins have large leaves and long, rambling vines that often invade other areas of the garden.

Maintenance

Squash and pumpkins are heavy feeders and respond well to rich soils that contain ample composted manure. The soil should be well drained. After they have germinated, thin seedlings to the strongest three plants per hill. For rows, thin to stand an average of 1 foot apart, more for larger pumpkins and winter squash.

A side dressing of fertilizer and regular feedings of fertilizer will significantly help the health of the plant as well as the size of the harvest.

Mulching is very beneficial for squash because it helps the soil to retain needed moisture as well as keeping weeds and insect problems down. A good layer of mulch will also help to keep fruits from rotting as they set on the ground.

Insect pests and diseases

Cucumber beetles, squash bugs, squash vine borers, mites, and aphids are the most common insect pests affecting squash and pumpkins. Cucumber beetles are oblong beetles that are yellowish-green in color, with three vertical black stripes down their backs. Rotenone gives good control for cucumber beetles. Cucumber beetles spread bacterial wilt and need to be controlled.

Adult squash bugs are rather large, a little over ½ inch long, winged, brownish black, and are sometimes mottled with gray or light brown, and flat-backed. Eggs are yellowish-brown to brick red laid in groups or clusters. Squash bugs tend to give off a disagreeable odor when crushed.

Both nymphs and adults suck sap from the leaves and stems of squash and related plants, apparently at the same time injecting a toxic substance into the plant causing a wilting known as Anasa wilt, named for the insects scientific name, Anasa Nistis. After wilting, vines and leaves turn black and crisp, and become brittle. Small plants are killed entirely, while larger plants will have one or several runners affected.

Squash bugs are often found in large populations, congregated in dense clusters on vines and unripe fruits. Sometimes no fruits are formed.

Bacterial wilt is a bacteria that invades the vascular, or water conducting, tissues of cucurbits. It causes a rapid wilting of the plant. Cucumbers and melons are more often attacked then are squash and pumpkins. These are attacked only occasionally. Progressive wilting occurs, beginning with a single leaf and spreading to including the entire plant. The bacterial organism produces a sticky substance that blocks the water flow through the vascular tissues, producing the typical wilt symptom. Plant wilting and stringing of sap when affected stems are cut are diagnostic for bacterial wilt. The bacterial wilt organism is carried from plant to plant by the striped cucumber beetle.

Prevent bacterial wilt by controlling the cucumber beetle. Begin beetle control early, as cucumber beetles may attack as soon as seedlings emerge from the soil. Use carbaryl, or diazinon for control. Frequent light applications of pesticides are best, beginning when the plants are young to achieve beetle control as well as to avoid plant damage from the insecticide. Promptly pull and destroy any diseased plants.

Anthracnose is a severe fungal disease that spreads rapidly in warm, wet weather. The first symptoms usually appear on older squash plant leaves as small yellowish circular spots. The disease spreads from older leaves to younger leaves. In warm, wet weather all the leaves may be attacked in rapid succession, giving the planting a "burned-out" appearance. Stems are also attacked, and light brown to black streaks develop. Circular, sunken, water-soaked spots develop on the fruits. These spots turn dark green to brown. In wet weather a pinkish ooze is produced within the spots. These pinkish spots are the fungus spores, which function like seeds. The spores are spread from plant to plant by running water, including splashing rain and by individuals working in the field when the vines are wet.

The anthracnose fungus overwinters on seed and on diseased crop refuse. Anthracnose control is difficult once the disease becomes widespread and serious. Prevention is the best control, including use of good quality seed, good sanitation practices, and appropriate crop rotation.

Powdery mildew is a common fungus disease of all squash, pumpkins and their relatives. Powdery white spots appear on the upper surfaces of older leaves, usually beginning around mid-season or later. During hot, humid weather the disease can progress rapidly, with the upper surfaces of all leaves developing a white powdery appearance. Severe powdery mildew outbreaks cause the leaves to turn yellow and wither. Planting with ample air space between vines will help prevent powdery mildew from taking hold of your plants. Pulling and destroying advanced-diseased plants, and practicing good garden sanitation will also help prevent this disease. There are several fungicides that can be purchased to help combat powdery mildew. Always read the label to be sure the fungicide is affective against this disease.

Mosaic viruses produce a patchwork or mosaic pattern of light and dark designs on the leaves and fruits of most cucurbits, including squash and pumpkins. Leaves become small and puckered, and the plants are usually severely stunted. Fruits develop knobs or warts on them, and often the fruits are misshapen.

Cucumber mosaic is very common. In addition to the mosaic pattern the edges of the leaves turn down, and the knobs on the fruits are light yellow. The cucumber mosaic virus is transmitted from plant to plant by aphids. Cucumber mosaic is readily spread from plant to plant on the hands of homesteaders working in the squash patch as well as by aphids.

Squash mosaic, which is caused by the squash mosaic virus, is transmitted from plant to plant by cucumber beetles, another reason to control this insect pest. The virus infects squash, cucumber, melon, and occasionally watermelon. The virus is also sometimes seed-borne.

Mosaic diseases are managed by using good quality seed and by controlling aphids and cucumber beetles throughout the season. Carbaryl is a good, low toxicity insecticide to use to control these beetles and aphids.

Begin insect control as soon as seedlings emerge from the soil. Do not plant squash bordering woods, shrubby areas, or weedy areas. Controlling all weeds, especially perennial weeds, around your squash plants will also help deter aphids. Pull and destroy diseased plants as soon as mosaic virus appears. Good sanitation will help to reduce virus spread. After handling any diseased plants, wash your hands well with detergent and water to reduce the risk of spreading the disease further.

Harvesting squash and pumpkins

Harvest summer squash by cutting the desired fruits off the vine with a knife. Summer squash should be harvested when still small, unless the fruits are being used for pickles or relish. You should be able to easily pierce the skin with a fingernail.

Elongated squash like zucchini and yellow squash should be picked when they are anywhere from 1 ½ to 3 inches in diameter.

Scallop type summer squash such as patty pan should be picked when they are 3 to 4 inches in diameter.

Remember, keeping your summer squash picked will encourage continuous production.

Winter squash varieties should be left on the vines until their rinds are hard. You can leave the squash on the vines until the vines die back, however, harvest all squash before frost hits.

When cutting winter squash and pumpkins from the vine, be sure to keep a couple of inches of stem attached to the fruit. Fruits without a stem are subject to faster rot, and should therefore be used first.

Cure squash and pumpkins in the sun or in a warm, ventilated area for a couple of weeks before bringing them into storage. Store them in a cool dry place after they are cured.

Winter squash varieties vary in size from an average of 15 lbs for hubbard to the 20 or more lbs for jumbo pink banana squash. There are the smaller, often only 1 lb, acorn squash, and the 3 to 5 lb butternut.

Pumpkins can be small, like the sugar pie pumpkins, or giants, like the Connecticut field pumpkins.

Summer squash, winter squash, and pumpkin can all be eaten fresh, frozen, dehydrated or canned.

Sweet potatoes

Sweet potatoes thrive in the long, hot summers of the South, however, they can be grown successfully wherever the growing season is of sufficient length, generally 120 frost-free days. Sweet potatoes are a low maintenance vegetable.

Planting guidelines

Sweet potatoes are the roots, or tubers, of a vine that is raised from the sprouts, or slips, of a parent plant.

One potato suspended on toothpicks in a container and half covered with water will produce several sprouts. Larger quantities can be grown by placing several sweet potatoes on a bed of sand and covering them with a 2 inch layer of moist sand or soil. Keep the sprouting tubers at about 75 degrees F. You can also purchase started slips at a nursery or garden center.

Start your slips about a month before warm weather settles in. During this month that the slips are developing they will grow up to 10 inches long and each will have several leaves. Remove the slips for planting by giving them a twist to separate them from the parent root.

Prepare the garden bed by mixing well composted manure into the top soil in the sweet potato bed. Using a hoe, create a long mound of the soil/manure mixture that is approximately 6 inches high and 1 foot wide. Flatten the top of this mound slightly.

Plant the slips 15 inches apart in the center of the mound, and set them about 6 inches into the ground, leaving at least two leaves aboveground. Water the slips in well

Maintenance

Sweet potatoes require very little care. A bit of weeding, done carefully so as not to injure the shallow roots, is usually all that is required.

Insect pests and diseases

Sweet potato weevils feed on the leaves of the vines, and the larva tunnel into the tuber. Keep the ground clear of leaves, weeds, and other garden debris to keep this pest under control.

Harvesting sweet potatoes

When sweet potato plants turn black after the first frost, the sweet potatoes are ready to harvest. In frost-free areas, sweet potatoes can be harvested after four months.

Dig sweet potatoes carefully, because their skins bruise easily. Let the tubers dry for several hours, and then spread them in a newspaper-lined boxes. Leave them in a dry, warm area for about two weeks to cure. Then, store in a cool, dry place.

Sweet potatoes can be canned or frozen after cooking, although the keeping quality of properly cured, undamaged tubers makes other forms of preservation unnecessary.

Tomatoes

Tomatoes are the most popular vegetable for home gardeners. No other vegetable comes close to it's popularity. Tomatoes come in a variety of sizes and colors. Generally yellow tomatoes are less acidic than the red ones are. There are grape tomatoes, cherry tomatoes, canning tomatoes, paste tomatoes, and slicing tomatoes, just to name a few!

Tomatoes are warm weather plants that thrive in full sun.

Planting guidelines

Tomatoes are usually started indoors. Sow two tomato seeds in each individual peat pot about eight to ten weeks before the last spring frost date for your area. Water the seeds in well and place them in a sunny window to germinate.

When the seedlings develop their first true leaves, thin to the strongest plant per pot. Just prior to planting them in your garden, after all danger of frost is passed, harden them off by bringing them outside during the daytime, for increasing hours, until you are leaving them out overnight. Use of a cold frame is recommended, but not a requirement if you don't have one. If frost is predicted, simply bring them indoors.

On planting day, pour liberal amounts of water, mixed with a water-soluble liquid fertilizer, onto each of them. Transplant them into the garden carefully. To minimize transplant shock, avoid disturbing the roots. This is why the use of peat pots is recommended, as the entire pot can be planted. Space plants about 24 inches apart, in rows that are set 3 feet apart.

Maintenance

Be sure to fertilize your tomato plants on a regular basis. Early applications should be higher in nitrogen, however, as blossoming occurs, switch to fertilizers which are higher in Phosphorus and Potassium to ensure fruit set. Too much Nitrogen fertilizer results in lots of lush green leaves, and little or no fruit.

Keep your tomato plants well watered. Deep watering is preferable, over more frequent, light watering. You want moisture to go deep to all the roots of the plant. Water directly to the roots. Be sure to keep water off the leaves if at all possible. Tomatoes are susceptible to plant disease that grows in wet, humid conditions.

To maximize your crop, and minimize disease and insect damage, stake or cage your tomatoes. They will reward you with more tomatoes, less rot, and the tomatoes will be cleaner because they will not be sitting on the ground. Harvesting is easier too, with fewer broken branches.

Insect pests and disease

Cutworms can be a problem to newly transplanted, young tomato plants. Cutworms are medium sized larvae of a family of moths. These larvae, or caterpillars, feed at night on the stems and roots of young plants, often cutting them off near the surface of the ground. They hide in soil by day.

If you have seedlings that are apparently cut near the soil surface, dig down in the soil around the root that remains in the soil. If it is a cutworm the culprit will most likely still be in the soil. Always handpick cutworms and destroy them.

Where cutworms are a problem, shields can be made by cutting down a paper tube such as paper towels come on, cutting it in half, and putting it in the ground around the newly transplanted plant. Be sure to bury about 3 inches of the tube to give complete protection.

Both tomato and tobacco hornworms attack tomato plants, causing damage to foliage and occasionally completely defoliating a plant in a matter of days. These caterpillars will also, on occasion, eat green fruits. Each of these large green caterpillars has a sharp spine, or horn, on its back. Hand pick them when you see them. Dusting with carbaryl will control these caterpillars. Bt can be applied as a preventative.

If you see one of these caterpillars covered with white eggs, do not destroy it. The eggs are those of a parasitic wasp that is very beneficial to your garden. Remove the caterpillar from your tomato plant, but don't destroy it. Put it aside where the wasp larvae can hatch and develop.

Slugs and snails will occasionally chew on the ripening fruit causing considerable damage. Keeping fruits off the ground by staking or caging, and cleaning up garden debris will help deter these pests.

Tiny black flea beetles can be controlled by using rotenone.

It is a fact that tomato plants emit a mild toxin that discourages many small insects from bothering them.

A number of problems can arise in tomatoes, usually in mid summer heat and humidity.

Blights and fungus infections can occur in the high humidity. Early treatment with fungicides is effective. Spacing plants too close cuts down air circulation and promotes fungal diseases.

Most tomato wilts can be prevented by planting resistant varieties.

Blossom end rot can also affect the fruit. This is a round, brown, indented spot on the bottom of the tomato. It is caused by either uneven watering or a lack of calcium in the soil. The spot may enlarge until it covers as much as 1/3 to one-half of the entire fruit surface, or the spot may remain small and superficial. Large lesions soon dry out and become flattened, black, and leathery in appearance and texture.

Another common predisposing factor is cultivating too close to the plant. This practice destroys or damages the roots, hindering their ability to take up water and minerals. Tomatoes planted in cold, heavy soils often have poorly developed root systems and succumb to this condition as well.

Control of blossom end rot is dependent upon maintaining adequate supplies of moisture and calcium to the developing fruits. Tomatoes should be planted in well-drained, loamy soils. Planting tomatoes in warmer soils also helps to alleviate the problem. Irrigation must be sufficient to maintain a steady even growth rate of the plants. Mulching of the soil is often helpful in maintaining adequate soil moisture, especially in times of drought. When cultivation is necessary, it should not be too near the plants nor too deep, so as to avoid damaging the tender roots. Shading the plants is often helpful when hot, dry winds are blowing, and soil moisture is low. Use of fertilizer low in nitrogen, but high in superphosphate, such as 4–12–4 or 5–20–5, will do much to alleviate the problem of blossom end rot. This disease does not spread from plant to plant in the field,

Do not water tomatoes at night in hot and humid weather if possible. Moisture and humidity combined with high temperatures promotes plant diseases. Whenever possible, water tomatoes at the roots.

Harvesting tomatoes

When the fruits begin to turn red or

yellow, depending on the variety, begin checking the plants daily, and pick the tomatoes as they become fully ripe, but firm. Over-ripe tomatoes will fall off the vines and will rot quickly.

When a frost is expected you can try to cover your tomato plants to protect the remaining from the weather. You can also pick the green fruits and ripen them indoors in a warm place or use them green. Green tomatoes can also be individually wrapped in newspaper and stored for several weeks in a cool, dark place.

Tomatoes can be eaten fresh, dried, canned, frozen, or made into any number of wonderful dishes.

Turnips and rutabagas

Turnips and rutabagas are mustards and are members of the cabbage family. They are cool season crops, and as such, they must be grown in the cooler temperatures of early spring and late fall. Turnips and rutabagas need full sun and a well drained soil for best crop production. Turnips are a dual purpose crop. The leaves are used for greens and the root is used in a similar manner to potatoes.

Turnips are easy to grow if sown in the proper season. Turnips generally mature in about 2 months, while rutabagas take 100 days or more to mature.

Planting guidelines

The soil should be raked smooth and free of debris. If the soil is heavy clay, add compost or other organic matter to loosen the soil. This is very important if you are growing your turnips for the roots because heavy soil can cause turnip roots to be rough and misshapen.

For summer use, turnips should be planted as early in the spring as possible. For fall harvest, plant rutabagas about 100 days before the first frost and plant turnips about 3 to 4 weeks later.

Dig the soil 10 to 12 inches deep. Work several pounds of well composted manure into the soil. Prepare a good seedbed, scatter the seed and rake the seed in lightly.

Sow seeds ½ inch deep, and about 1 inch apart in rows that are spaced 12 to 24 inches apart. Water if necessary to germinate the seed and establish the seedlings.

Thin rutabaga seedlings to six inches apart when they are two inches tall. Thin turnip seedlings to 2 to 4 inches apart when they are four inches tall. The removed plants are fine to use as greens.

Maintenance

There is generally no need to cultivate

turnips, however they do benefit from a good layer of mulch which will help the soil retain moisture as well as discouraging weed growth.

Insect pests and diseases

Turnips and rutabagas are attacked by two different flea beetles. These insects eat holes in the leaves, chew stems and cause extensive plant loss. The cabbage flea beetle and the striped flea beetle feed exclusively on members of the cabbage family, including related weeds such as yellow rocket. Both flea beetles can be controlled with carbaryl or rotenone.

Turnip crops can also be damaged by infestations of the common turnip aphid. This insect feeds on the undersides of the leaves and may be so close to the ground that it is difficult to reach with a dust or a spray. In cases of severe infestation, the outer leaves curl and turn yellow. Aphid-tolerant varieties can have numerous side shoots. Secondary, rot-producing bacteria can often gain entrance through these infections and cause even greater crop losses.

Good garden sanitation and proper crop rotation are give some protection against this insect.

Turnip crops may suffer from clubroot, root knot, leaf spot, white rust, scab, anthracnose, and turnip mosaic virus.

Club Root can be identified by stunted growth especially if infected as seedlings. Discolored leaves will eventually wilt. Roots are a thick distorted mass. Clubroot is caused by a fungus within the soil. The only treatment is prevention.

White rust is also caused by a fungus. This disease causes normal root development to be retarded, resulting in tough, woody roots that vital to controlling these diseases. Do not over-crowd turnips or rutabagas and never grow turnips or rutabagas in soil where any member of the cabbage family, or any other root crop has been grown for two years.

Resistant varieties are available for most of these diseases.

Harvest turnips and rutabagas

If you have planted turnips for greens only, harvest the tops as needed when they are 4 to 6 inches tall, or harvest the entire plant about a month after planting to process the tops for storage. If the growing points are not removed, tops continue to regrow.

If want to harvest roots as well as tops then you should not harvest all the tops from the plants. Cut only a few leaves off each plant, leaving the rest to nourish the plant.

Turnip roots are at their best when they are 2 to 3 inches in diameter. Older turnips tend to be tough and woody. Although turnips tend to get sweeter if they are allowed to get hit with a light frost, they must be harvested before any hard frost sets in.

Store turnips by burying them in moist sand in a root cellar.

Rutabagas are ready to eat when they reach a size of about 3 inches in diameter, however, they can be left to grow until much larger if desired. If left to get much beyond 5 inches in diameter they will be tough and woody.

Rutabagas keep very well in storage by placing them in moist sand in a root cellar. They can also be left in the garden, and covered with a thick layer of mulch and dug up as needed during the winter months.

CHAPTER THREE
The Herb Garden

ALL HOMESTEADERS SHOULD GROW their own herbs. Herbs are an important part of flavoring foods, making home remedies, and scenting soaps and candles.

Herbs can be grown in a sunny corner of the vegetable garden or they can be grown in their own decorative patch, usually just outside the kitchen door. Herbs can be worked into an edible landscaping theme with ease because most herbs are attractive as well as useful.

When planning your herb garden, bear in mind the ultimate size and growing habits of the herbs you want to plant.

Some herbs, such as rosemary, are tall perennials that require full sun and adequate moisture. Others, like the thymes are lower growing and sprawling.

Plan the garden so that the taller

plants are in the back where they won't cast shade on smaller plants. Mints should be planted in an area where they can spread without overwhelming other desirable plants.

Herbs can also be planted in the vegetable garden as companion plants. Many herbs, such as rosemary, garlic, mint, parsley, and sage have a tendency to repel certain insects, making them very desirable as companion plants to more vulnerable vegetables. Other herbs, such as hyssop, balm, dill, and thyme are among the herbs that attract bees, thereby increasing pollination.

Some herbs help to either increase, or inhibit, the growth and development of vegetable plants. For example, green beans are improved by having summer savory nearby, as savory dispels bean beetles, while those same beans are inhibited by having any of the alliums, such as chives or garlic growing nearby. Dill is a good companion for the cabbage family, while it is inhibitive of the growth of carrot roots.

Fennel does not make a good companion plant for any garden vegetables, so avoid placing it in close proximity to any vegetable.

Horseradish will repel potato beetles in the garden, however, horseradish must be contained somehow or it will spread rampantly throughout the garden. Similarly, the mints repel aphids in the garden but will take over if allowed to. Keep mint planted in pots and place the pots around your tomatoes and cabbages.

Nasturtiums are wonderful natural insect repellents if planted among radishes, squash, melons, cucumbers and cabbages. Nasturtiums will repel aphids, squash bugs, and pumpkin beetles.

Sage dispels cabbage moths, flea beetles, and slugs on cabbages and carrots, however it dislikes cucumbers. Thyme dispels cabbage worms when planted as a companion to any of the cabbage family.

Most herbs can be grown out in the garden in the summer months and then transplanted into pots and brought indoors during the winter months. Any herbs grown in the house over the winter will need to be placed in a bright, sunny window to maintain health.

Perennial herbs such as marjoram, chives, mints and winter savory can be grown from cuttings or divisions taken in the fall. Basil, dill, parsley, and other annual herbs can be started from seeds sown outdoors and transplanted into pots for winter growth, or they can be directly sown into pots to remain indoors throughout the colder weather.

There are many herbs available for the homestead gardener today.

Angelica

Angelica is a tall border plant that is spectacular, especially at maturity. Angelica usually bears clusters of greenish flowers in late spring of the second year, and reaches a height of from 4 to 7 feet and a spread of 3 feet. The crumbled or chopped leaves of angelica can replace some or all of the sugar in fruit pies. Seeds can be brewed into a semi-sweet tea.

Once established, angelica will self-sow if allowed to set seeds and die back as a biennial.

Angelica will also attract beneficial insects to your garden.

Sow in early summer in partial shade in moist, rich soil. Sow by planting groups of 3 seeds about 2 feet apart. When seedlings have 3 leaves, thin each group to leave the strongest plant.

Angelica can be grown as either a biennial or a perennial. If you remove the flower heads the plants will live as perennials. If you leave the flower heads on the plants, it will die back after flowering. Keeping the herb garden weeded and mulched will create healthy plants and increased yield.

Angelica is not usually bothered by

insects or disease organisms.

To harvest angelica, cut young stems in the spring of the second year, before the flower heads develop. Leaves can be harvested all summer long. Harvest seeds when they are ripe in late summer or early fall.

Anise

Anise is known for it's delightful little licorice-flavored seeds called aniseed. Anise is used in cough remedies and anti-itch ointments. The seed adds flavor to baked goods, candies, and applesauce, as well as many international dishes. The fresh leaves can be used in salads.

Anise is a hardy annual that grows about 18 inches tall with a spread of about 12 inches. Anise prefers full sun and rich, well drained soils.

Anise is grown from seeds. Sow seeds in mid-spring about ½ inch deep and 4 inches apart. When seedlings are about 2 inches tall, thin to stand about 12 inches apart.

Harvest aniseed when the seeds are ripe, but have not yet fallen from the plant. Cut the flower heads into a paper bag to prevent seed loss. Then, thresh by hand over a large sheet of white paper. Allow the seeds to dry and store them in an airtight container in a cool, dry place.

Balm

This herb is often called lemon balm because of the delightful, lemony fragrance and flavor of its light green leaves. Balm bears small white or pale yellow flowers late in the summer and early fall. These flowers are highly attractive to bees.

The leaves of balm lends their gentle lemony flavor to teas, puddings, soups, fruit drinks, and can be used to replace some of the sugar in fruit pies.

Balm is a hardy perennial that grows to about 4 feet in height and has a spread of 18 inches or more. Balm thrives in full sun or partial shade, and will grow in any soil having adequate drainage.

Sow several of the tiny seeds in individual peat pots in early spring. Thin the tiny seedlings to the strongest plant per pot. When the seedlings reach 4 inches in height, and after all danger of frost has passed in the spring, they can be transplanted into the herb garden. Set them out 1 foot apart in all directions.

Balm can also be cultivated by dividing the root clumps into several pieces in the spring. Be sure to keep 3 or 4 buds on each root clump. Plant these root clumps 12 inches apart in rows that are spaced 18 inches apart.

Cut shoots individually as soon as

flowers appear, continuing to cut shoots as needed until fall. Tie a bunch of shoots together, and hang upside down in a warm, dry place to fully dry and allow the essential oils to travel downward and into the leaves. Remove leaves from stems when completely dry, and store leaves in an airtight container in a dry place.

Basil

Basil is grown as a perennial in the tropical regions, and is considered an annual in temperate climates.

Basil has shiny green leaves that have a sweet, almost clove-like scent and upright growing habits. It is frequently grown as a pot herb in the kitchen as well as in the garden. Basil develops white or purplish flower spikes late in summer.

There are numerous varieties of basil to choose from. Plant basil among your tomatoes helps repel hornworms.

Sow basil seeds directly into the garden late in the spring when all danger of frost has passed and the weather has warmed.

Basil can fall prey to numerous fungi in cool soil. Whether you sow seeds or set in transplants, make sure the ground has warmed thoroughly. Sow the seed about 2 inches apart in rows that are about 1 foot apart. Thin seedlings to stand about 6 to 9 inches apart. You can also start seeds indoors four to six weeks before the last expected frost and transplant them into the garden later. Sow 2 seeds per individual peat pot, water well and set in a sunny window to germinate. When seedlings are 2 inches tall, thin to the strongest plant per pot.

When the plants are about 6 inches tall, pinch off the tips to encourage bushier growth. Remove flower buds as soon as they appear, to maintain plant development. Pick fresh, young basil leaves whenever you need them. The more leaves you harvest, the more the plant will produce.

Let some flowers remain on a few plants if you live in a warm climate and want your basil to self-sow.

In warmer climates you can harvest basil by cutting the plant down to about 2 inches from the ground, leaving 2 or 3 leaves on the plant. In more northern climates, pull the entire plant out in the fall, then cut off and discard the root. Tie the stems together in a bunch and hang them upside down in a warm, dry place to dry completely. Hanging upside down sends the essential oils from the stems into the leaves.

Dried leaves can then be stored in an airtight container. Basil can also be frozen fresh or kept refrigerated in jars with alternate layers of basil leaves and salt, and topped with olive oil.

Borage

Borage is a hardy annual grown for both its leaves and its pleasantly scented blue, lavender, or pink flowers. Borage prefers full sun or partial shade and will grow in any adequately drained soil. Borage helps repel tomato worms when planted near tomato plants. Borage is attractive to bees and is a good companion for strawberries and other fruits.

Grown from seeds, borage will self seed if the flowers are left on the plant. Plant borage in late fall or early spring. Sow directly in the garden after the weather has warmed and all danger of frost is passed.

Borage can also be started indoors about 4 weeks before the last expected spring frost date. Sow seeds in individual peat pots for easy transplant to the garden. Thin seedlings to the strongest plant per pot when seedlings reach about 2 inches tall. Transplant outdoors after all danger of frost has passed in the spring.

Full grown, borage reaches a height of from 1 to 3 feet and a spread of 1 foot.

The first leaves can be harvested about 6 weeks after seed germination.

Pick flower buds as they appear, just before they open. You can use the flowers, and by removing them from the plants, you are encouraging more vigorous plant growth.

Harvest young leaves to use in salads for their cool, cucumber-like flavor. Candy the delicate flowers for cake or pastry decoration or float them in punch. Borage is also used in many herbal home remedies.

Caraway

Caraway is a hardy biennial whose leaves and greenish-white flower heads resemble those of carrots. It prefers full sun and will thrive in any well drained soil. Caraway is grown for both its leaves and seeds.

Caraway attains a height of about 2 feet and a spread of 9 to 12 inches at maturity. Caraway is grown from seeds and will self sow. Plant seed in late spring or early summer directly in the garden, or start them indoors in individual peat pots. Caraway develops long tap roots so it is important to use individual peat pots to avoid damaging these roots during transplanting. Sow 2 seeds per pot, water well and set in a sunny window to germinate. When seedlings reach about 1 inch in height, thin seedlings to the strongest plant per pot. Transplant to the garden when all danger of frost is passed and the weather has warmed sufficiently.

The first year will see only the carrot-like greenery appear. During the summer of the second year the white flower umbrels form. After the crescent shaped, brown seeds mature the entire plant will die.

Harvest seeds into a paper bag for cleaner harvest, or harvest seed heads before they are fully ripe and hang the heads upside down in a warm, dry place to finish ripening. Set trays or a sheet under the ripening heads to catch any seeds that fall. Store dried seeds in airtight jars.

Caraway seeds have been used to aid digestion and in various other home remedies.

Caraway seeds add tangy flavor to baked goods, roasted meats, and apple dishes. A few seeds tied in a spice bag and added to cooking water will dispel the odor of the cooking cabbage.

Young caraway leaves can be used in salads while older leaves are cooked like spinach.

Chervil

Chervil, a hardy annual, is also known as French parsley. This plant's fernlike foliage reaches a mature height of about 2 feet and a spread of up to 1 foot.

Chervil prefers partial shade and requires moist but well-drained soil. To encourage leaf production, pinch out the flower buds as they develop.

Chervil is best used fresh, but can be dried if desired. Fresh leaves are used in similar fashion to fresh parsley, and has a slight anise flavor. Chervil is a versatile herb that is used in egg dishes, and in soups and sauces. It is also used in fish, poultry, and meat dishes.

Chervil is cultivated by seed and can be direct sown into the herb garden after all danger of frost has passed in the spring. Sow seeds ¼ inch deep and 4 or 5 inches apart in rows that are about 1 foot apart. When seedlings are about 2 inches tall, thin to stand about 9 inches apart. Chervil can also be grown indoors as a houseplant for continuous year-round supply of the fresh herb.

You can begin to harvest fresh leaves around 6 to 8 weeks after sowing. Cut plant back to the ground to harvest the entire plant. If drying chervil, hang plant upside down in a warm, dry spot for a few weeks, or until sufficiently dry. Drying upside down allows the essential oils to flow down from the roots and stems and into the leaves.

Store dried chervil in an airtight container in a cool, dry place.

Chives

Chives are hardy perennials that form long, thin, tubular foliage that have a mild onion flavor. Mature plants form purple, pom-pom flower heads atop thin stalks. Chives usually flower in mid summer. With a height of 6 to 10 inches and a spread of 12 inches, chives make excellent border plants. They are excellent companion plants for carrots because of their ability to discourage a fungal disease that attacks carrots.

Chives are propagated by seeds and by root divisions. These plants prefer full sun but will tolerate slight shade, and will grow in any well-drained soil. Sow seeds in the spring or fall, about ½ inch deep, one seed every 3 inches, in rows that are set about 1 foot apart. As soon as seedlings are established, thin to 6 inches apart. Use the thinning in the salad bowl for a delicious treat.

Nursery grown plants can be purchased for immediate transplant into the garden. For root divisions, lift established clumps of chives and divide them into 2 or 3 smaller clumps every 3 or 4 years in the spring.

Chives can be harvested 4 to 6 months

after sowing, earlier if nursery grown plants are used. Cut the long, slender leaves off close to the ground.

Chop chives, or snip them with scissors into small pieces, and add them to green salads, chicken salad, egg dishes, sandwich spreads, potato dishes, and sauces.

Chives do not dry well, the leaves lose color and flavor upon drying. Instead, bring a clump indoors to grow as a houseplant for fresh chives all winter. Chives can also be successfully preserved by freezing.

Garlic chives are another delightful relative that lend a garlicky flavor to salads, stir fries, and potato dishes. Grown the same way as regular chives, they are only slightly less hardy. They form 2 inch wide, white flower heads that have a sweet scent. You'll want to plant both varieties in your garden!

Coriander

Coriander, a hardy annual that grows 2 to 3 feet tall, has parsley-like leaves and rosy white flowers. The flowers have a somewhat unpleasant smell. Coriander is grown for both the leaves, which are also known as cilantro, as well as for the fragrant seeds.

Coriander prefers full sun and will

grow in any soil with adequate drainage. It reaches a height of about 18 inches and a spread of 6 to 9 inches.

Coriander is grown from seeds that are sown directly in the garden early in the spring. Sow seeds ¼ inch deep with seeds spaced 3 inches apart. Rows should be spaced about 12 inches apart. Thin established seedlings to stand 6 inches apart.

Coriander seeds are dried and then ground to a powder. Ground coriander seed is used in veal and pork recipes as well as in baked goods, sausages, stews and soups. Young coriander leaves are an important herb in Mexican cuisine, some salads, soups and sandwich spreads.

Harvest coriander leaves as needed once plants are of sufficient size. Coriander leaves are best used fresh.

Harvest seed by cutting the seed heads when they ripen. Spread seed heads on trays to dry in the sun or in a dehydrator. Thresh the seeds by hand over white paper. Store in airtight jars when fully dry.

Dill

Light green, feathery foliage rises 2 to 3 feet tall and spreads 9 to 12 inches wide. Dill prefers full sun and moist, but well-drained soil. Dill is highly aromatic and is known for attracting honeybees. It develops umbrella-like blossoms in midsummer. Dill makes a good companion crop for the cabbage family, but does not like carrots.

Dill is easily grown from seeds sown in early spring. Sow seeds ¼ inch deep and about 4 inches apart. Keep rows spaced about 9 inches apart. Thin established seedlings to stand about 9 inches apart.

Both the seeds and the leaves of dill are used. The leaves, known as dillweed, can be used fresh, or dried for future use. Dillweed is used to flavor pickles and relishes, in fish dishes, omelets, potatoes, soups and salads. A head of dill seed makes a delightful addition to a jar of dill pickles.

Dill seeds are used in similar fashion, but one should bear in mind that the seeds are stronger flavored than the leaves.

Dill leaves can be harvested just as the flowers open. Dillweed can be dried by hanging the stems upside down in a warm, dry place until the stems and leaves are dry. When the leaves and stems are sufficiently dry, the dillweed can be stored in airtight containers in a cool, dry place.

Cut flower stems in dry weather when the seed heads are sufficiently ripe. Hang seed heads upside down in a sunny, dry location, over white paper to dry. The paper will catch any seeds that may fall. Hand thresh the seeds and store them in an airtight container in a cool, dry room.

Fennel

Fennel is a perennial herb that grows 3 to 4 feet tall with a spread of about 2 feet. It has feathery foliage similar to that of dill, however, fennel foliage can be either green or copper colored. Fennel prefers full sun and will grow in any adequately drained soil.

Sow fennel in mid-spring, after all danger of frost has passed. Sow seeds in groups of 3 or 4 seeds about 18 inches apart. When seedlings are established, thin to the strongest plant in each group. Although considered a perennial, you may wish to replant every 2 or 3 years if your plants are heavily harvested.

Fennel leaves have a sweet flavor that is often used in fish, veal or pork dishes, soups and salads. Fennel seeds have a more pungent flavor and are used in sausages, sauerkraut, and on baked goods. Florence fennel leaves are cooked and eaten as a vegetable. Sicilian fennel stems are blanched and used like celery.

Harvest the fresh leaves and stems as needed before the flowers bloom. Leaves can be dried, or frozen for winter use. If fennel is grown for seeds, cut the stems in autumn and hang the mature flower heads upside down in a warm, dry spot to dry. Be sure to place a sheet of paper beneath them to catch any seeds that drop.

Store the dried seeds in an airtight container in a cool, dry place.

Horseradish

This perennial member of the mustard family develops invasive and voracious roots will quickly choke out all other plants if left to spread. Leaves are large and coarse, growing to a height of 2 to 3 feet with a spread of 12 to 18 inches. Horseradish prefers either full sun or partial shade and thrives in deep, moist soil.

Horseradish discourages potato fungal diseases, as well as blister beetles, when planted near potatoes. Use care to keep the horseradish roots well contained to avoid invasion into the potato patch.

Horseradish is propagated by root cuttings taken in early spring. Using a shovel, dig up a couple of larger roots and cut them into 3 inch pieces. Plant these rootlets in pots containing equal parts of a good gardening mix and sand. Make holes in the potted mixture 2 inches apart and 3 inches deep. Insert rootlets vertically in holes and cover with soil mixture. Water well and keep moist, but not soaking.

When cuttings have developed two or three sets of leaves, transplant them into the garden. Horseradish rootlets can also be planted in 3 inch deep holes directly in the prepared horseradish patch.

Horseradish root is shredded fine to make a popular spicy, hot sauce for beef, fish, game and other meat dishes.

Harvest horseradish by lifting roots in late autumn. Be careful not to leave any broken root pieces behind unless you don't care if the roots spread. Trim away and discard small side roots.

Horseradish can be ground up and mixed with vinegar and bottled, or the roots can be stored whole in the refrigerator for short term storage, or placed in moist sand in the root cellar for longer term storage. Younger roots are often preferred for culinary purposes.

Marjoram

Marjoram is a valuable garden plant for it's decorative and fragrant foliage, attractive flowers, and range of uses. There are several varieties available, from the smaller, hardier pot marjoram, with a mature height and spread of about 1 foot, to the larger, milder sweet marjoram that attains a mature height of 2 feet and a spread of 18 inches. Marjoram is a tender perennial that requires some shelter from the cold winters of more northern climates. It prefers full sun and thrives in warm, rich, well drained soil.

Marjoram is grown from seeds or from plant divisions. Sow seeds 6 inches apart in rich soil early in the spring after all danger of frost has passed. Thin established seedlings to stand 1 foot apart. Plants can also be grow from plant divisions in mid-spring. Simply cut a large clump in half or thirds and replant in the garden or in pots. Water well.

In Northern areas it may be necessary to pot marjoram and bring it indoors or into the greenhouse for the winter, returning it to the herb garden in the spring after the weather has warmed and all danger of frost has passed.

Fresh or dried marjoram leaves are used in lamb, pork, and veal dishes, as well as in stews, soups, stuffing's, sauces, as well as in egg and cheese dishes. Marjoram is also used in herbal teas, sachets, and several home remedies.

Harvest fresh leaves as needed before the flowers open in midsummer. Encourage leaf production by picking off flower heads as they appear.

Marjoram can be dried by hanging

bunches of cut stems upside down in a warm, dry room. Store dried, crushed leaves in an airtight container.

Mint

There are countless varieties of mint grown to choose from. Some species grow wild and all of them are very desirable garden plants. The most popular of the mints are apple mint, peppermint, and spearmint, however, there are also pineapple mint, chocolate mint, catnip, and others.

The white or purple flower spikes are attractive, but pinching them off as they appear helps to encourage leaf growth. Mints are perennial and reach a mature height of 2 to 3 feet and a spread of 18 to 24 inches. They prefer partial shade, in rich, moist, well-drained soil.

Mints are repellent to white cabbage butterflies and make good companions for the cabbage family. They also repel ants and therefore help protect plants from aphid infestations. Use care when inter-planting mints with other plants in your garden.

Mint can be invasive and should be confined to its own spot or planted in pots and scattered about the garden.

Mints are grown from root divisions. In autumn or spring, plant 4 to 6 inch pieces of root 2 inches deep and 12 inches apart. Water well.

To harvest mint, pick leaves as needed. For large harvest, cut plants even with the ground in midsummer. Mint leaves can be used fresh, or dried. To dry mint, hang cut plants upside down in a warm, dry spot. Store dried leaves in an airtight container in a cool, dry place.

The roots left behind in the herb garden will send up new foliage for a second crop.

Oregano

Oregano is also called wild marjoram. It is similar to sweet marjoram, but is shrubbier. Oregano has a sharper fragrance and flavor. Roots tend to be mildly invasive of neighboring plants.

A hardy perennial with a mature height of about 2 feet and a spread of 18 to 24 inches, oregano prefers full sun and will grow in any well-drained soil.

Oregano is grown from seeds or from root divisions. To grow from seeds, sow the seed about 6 inches apart in spring or autumn. Thin established seedlings to 12 inches apart. Dig clumps of oregano in mid-spring and divide in 2 or 3 pieces. Replant in the herb garden and water well.

Oregano can be used fresh or dried. It

is often used in Italian, Spanish, and Mexican cooking, especially in meat and tomato dishes. Oregano is also used in salads, stews, fish dishes, and in egg and cheese dishes. Oregano is also used in many herbal remedy recipes.

Harvest fresh oregano leaves as needed. For drying, cut off the top 6 inches of leaves just before the flowers open. Bunch and tie the stems together and hang upside down in a warm, dry place to dry thoroughly.

Store the dried and crumbled leaves in an airtight container.

Parsley

Parsley has been used since ancient days to sweeten the breath after eating dishes containing onions or garlic. Parsley is a biennial plant that is usually grown as an annual. It will thrive in either full sun, or partial shade and likes rich, moist soil. Parsley grows to a mature height of about 1 foot and will reach a spread of about 1 foot as well. Parsley is a good companion plant for tomatoes.

Parsley can be grown easily from seed. Soak seed overnight and broadcast thinly in the herb garden the next morning. Cover very lightly with soil. Water well. Thin established seedlings to stand about 10 inches apart. If parsley plants are left in the ground to grow throughout the second year, they will go to seed and self sow.

Parsley is sometimes attacked by swallowtail butterfly larva. These black and yellow striped caterpillars can be handpicked for control.

Parsley is best known as a garnish for foods, however it has real use as an herb in cooking. Mix leaves into salads, soups, stews, casseroles, and egg dishes.

Harvest parsley by cutting the stems as desired, however, no more then 2 or 3 stems from any one plant. If harvesting from plants the second year, harvest leaves before the plants flower or the leaves turn bitter.

Parsley leaves can be frozen or dried for future use. If you dry parsley, dip the leaves into boiling water and then place on a cookie sheet in the oven at about 400 degrees F for about 1 minute to dry. Store dried leaves in a cool, dry place in an airtight container.

Rosemary

Rosemary is a tender, perennial evergreen that is valuable as a landscape feature wherever the ground does not freeze solid in the winter. Its needle-like leaves have a somewhat piney scent. Rosemary is repellent to cabbage butterflies, carrot rust flies, and mosquitoes.

Rosemary reaches a mature height of from 2 to 6 feet tall, and a spread of from 2 to 6 feet wide. It prefers full sun or partial shade in light, well-drained soil.

Rosemary can be grown from seeds, however, buying started plants is more satisfactory. Sow seed in individual peat pots indoors in early spring. Water well, and place in a sunny window to germinate. Keep moisture level even during germination. Set out plants about 2 feet apart in late spring, after all danger of frost is passed.

Rosemary does wonderful things to pork roasts, veal, or poultry when sprinkled on before roasting. Sprinkle chopped leaves over beef or fish before grilling or broiling. Use sparingly in soups, sauces, stews, or in vegetable dishes. Rosemary is also known for the tasty tea it makes.

Harvest rosemary by cutting sprigs from the branches as needed. Rosemary can be used fresh, or it can be frozen or dried for future use. Dry rosemary by hanging bunches of sprigs upside down to dry in a warm, dry place. Crumble the dried sprigs to loosen the leaves from the stems.

Store dried rosemary in airtight containers in a cool, dry place.

Sage

A hardy perennial or evergreen sub shrub, Sage is good as a low background plant in the herb garden. Sage has narrow gray-green, pale yellow, or purplish leaves that reach a height of 2 feet and a spread of about 18 inches. Sage prefers full sun and will grow in any well-drained soil. Sage is a fairly drought resistant plant. Be sure to avoid over watering.

Sage is repellent to white cabbage butterflies, carrot flies, and ticks. However, do not grow sage near annual seedbeds, as it has a tendency to inhibit root development.

Sage can be grown from seeds sown in early spring in individual peat pots, and transplanted outdoors in the spring after all danger of frost has passed. Sow 2 seeds per pot and water well. Set in a sunny window to germinate. After seedlings are established, thin to the strongest plant per pot. Transplant to about 1 foot apart in the garden in spring.

Sage can also be grown from root divisions made in spring or early fall every 2 or 3 years. Dig the plant with a spade and gently cut the root into 2 or 3 pieces, replanting each piece within the herb garden.

Sage has been used as in herbal remedies since ancient times. Dried sage is a traditional ingredient in stuffings and dressings, as well as in sausages. It is delicious used with lamb, pork, and in cheese and egg dishes.

Harvest sage leaves as needed for fresh use. To dry sage, cut the top 6 inches of stalks, before the flowers open in early summer. Bundle and tie leaves and hang upside down to dry. Repeat harvesting and drying as desired as new growth develops.

Store dried sage leaves in airtight containers in a cool, dry place.

Savory, summer

Summer savory is an annual with small, aromatic leaves and tiny lavender or pink flowers covering the plant in midsummer. The blossoms are attractive to honey bees.

Summer savory makes a good companion plant for green beans and onions. With a height of 12 to 18 inches and a spread of only 6 to 12 inches, summer savory will fit into the smallest herb garden plan.

Summer savory prefers full sun and rich, light, well-drained soil.

Grown from seeds, summer savory can be direct seeded in the garden in mid spring after all danger of frost has passed. Broadcast the seed and cover very lightly with soil. Water well. Summer savory germinates slowly, so allow 4 weeks for germination. You can also start seeds indoors in early spring by sowing 2 seeds per individual peat pot. Water well, and set pots in a sunny window to germinate. Maintain even moisture throughout the germination period. Transplant into the garden after all danger of frost has passed in the spring. Thin established seedlings to stand 6 to 9 inches apart.

Summer savory leaves have a peppery flavor that is a traditional seasoning in bean dishes. It is also used in sausages, stuffings, meat pies, soups, stews, rice, and meat sauces. Add fresh leaves to salads, fish dishes, and egg dishes, or to vinegar for salad dressings.

Harvest summer savory leaves before flowers form in midsummer for the best flavor. Cut the entire plant partially back for a second crop. Hang bundle of leaves upside down in a warm, dry place until sufficiently dry.

Store dried summer savory leaves in an airtight container.

Savory, winter

Winter savory is a lower growing, and wider spreading plant than summer savory. Its glossy, dark green leaves less aromatic then summer savory, and the flowers are larger.

Winter savory is a hardy perennial that attains a mature height of 6 to 12 inches and a spread of from 12 to 18 inches. It prefers full sun, and sandy, well-drained soil.

Winter savory germinates slowly, so should be planted in the early spring, indoors in individual peat pots, or outdoors, directly in the garden, in the fall in warmer climates. Sow 2 seeds per peat pot, water well, and place in a sunny window to germinate. Maintain even soil moisture throughout the germination period. Thin established seedlings to the strongest plant per pot. Allow at least 4 weeks for germination. Transplant into the herb garden after all danger of frost has passed in the spring.

Set plants 12 inches apart in all directions.

Winter savory is used in fish dishes and meats, similar to the way summer savory is used, although winter savory has a stronger flavor. Use according to tastes. Combined with basil for a substitute for salt and pepper. Crushed fresh leaves rubbed on the skin make a good insect repellent.

Harvest winter savory by picking fresh leaves as needed. Cut plants halfway to the ground for a second crop. Dry leaves by bunching and tying and hanging upside down in a warm, dry place. Crumble leaves and store in an airtight container.

Tarragon

A hardy perennial, tarragon has fragrant green leaves on woody stems that reach a mature height of about 2 feet and a spread of about 15 inches. Tarragon prefers full sun and well-drained soil that is not too rich and is a little on the dry side.

Tarragon does not grow true from seed, therefore root divisions are the main form of propagation for the homestead gardener. Divide clumps of tarragon in early spring and transplant each division back into the herb garden. Protect the semi-tender roots from long freezes by heaping leaves, straw or other mulches on the over-wintering plants.

Tarragon's anise flavored leaves are used in salads, egg dishes, poultry, stews, soups, lamb, fish, vegetable sauces, and egg dishes. Tarragon is a primary herb for use in tartar sauce and flavored vinegars.

Harvest fresh tarragon leaves by picking as needed. Leaves can be frozen or dried. To dry leaves, bunch, tie, and hang leaves upside down in a warm, dry place.

Store in an airtight container when dry.

Cut plants down to the soil surface in

autumn before mulching for the winter months.

Thyme

Thyme, like mint, is available in a wide variety of colors, fragrances, and growing habits. All of them are hardy perennials that prefer full sun and rich, well-drained soil.

Thyme grows an average of 6 to 8 inches tall and has an average spread of 9 to 12 inches.

Most varieties of thyme do not grow true from seed. Common thyme can be grown from seed with some measure of success.

Sow common thyme in individual peat pots, 2 seeds per pot. Water well and set pots in a sunny window to germinate. When seedlings are established, thin to the strongest plant per pot. Transplant into the herb garden when seedlings are about 3 inches tall, setting plants 6 to 9 inches apart in rows set 1 foot apart.

Plant nursery grown stock in the spring after the soil is warmed and all danger of frost has passed.

Thyme can be propagated from root divisions as well. Dig up and divide established plants every 3 or 4 years. Some forms of thyme have a tendency to spread wherever branches touch the soil. These rootings can be cut off the main plant and transplanted.

Thyme is used to add flavor to meats before roasting as well as soups, stews and stuffings. Thyme is also used in many egg, cheese, and fish dishes. Lemon thyme is often used in desert dishes and with fruits.

To harvest thyme, pick leaves as needed. Cut plants for drying just before flowers open in early summer. Hang bunches of thyme upside down in a warm, dry place so that the essential oils will travel down toward the leafs. When sufficiently dry, crumble leaves and store in an airtight container in a cool, dry place.

CHAPTER FOUR
THE BERRY PATCH

JUICY STRAWBERRIES, PLUMP RED raspberries, sweet-tart blackberries, tangy gooseberries and glistening currants are just a few of the homegrown delights that any homesteader can enjoy.

Berry bushes and canes can easily fit into the plans for even the smallest of homesteads.

Blueberry bushes make wonderful hedges that are green in the spring, loaded with lovely and delicious blue fruits in the summer, and covered in eye appealing red foliage in the fall. Currants and gooseberries both do equally well in full sun or part shade and tolerate wetter soils then most other berries do.

Strawberries can be grown in pots on the patio or in pretty flower gardens near the back door. They can also be planted in pyramid style bed or in a conventional garden plot.

Most berries can be frozen, canned and dehydrated successfully for winter use.

Blackberries

Blackberries and raspberries are perennial plants that form canes. They can be erect, or trailing. Trailing types are known as loganberries, boysenberries, and dewberries. Blackberry roots live on indefinitely, but each year they send up canes that produce fruit the second season, then die. When selecting your bramble site, choose one that gets full sun. Blackberries will tolerate some shade, but the more sun they have the more fruit they'll produce, especially in cooler climates. Choose the site with care because the bramble bed will be a permanent bed.

Blackberries prefer rich, well drained soil with adequate moisture throughout the growing season.

Planting guidelines

Till the planting area well and dig in plenty of organic material or well-composted manure.

Buy plants from a reputable nursery or mail-order catalog, and make sure they're certified to be free of diseases and root nematodes, both of which can give your garden problems that you will have a terrible time getting rid of.

Plant in early spring in zone 5 through 1, and in fall or late winter in zone 6 through 11. Keep the plants moist until you can put them in the ground. Dig holes deep enough and wide enough to accommodate the root-ball, and moisten the soil. Set one blackberry plant in each hole at about the same level they were in their pots. Plant upright blackberries about four feet apart, and place trailing blackberries five to six feet apart. Keep rows about 10 feet apart. Cut blackberry canes back to six inches above the ground. After cutting the canes back, water them well. Apply a thick organic mulch such as straw mixed with rich leaf compost.

Maintenance

Give plants a constant supply of moisture, especially during fruiting. Replenish the layer of composted material every spring.

Planting thornless varieties makes working amongst your blackberries an easier chore. If you do plant varieties that have thorns you should protect yourself from painful pricks and stabs by wearing long pants, gloves, and long sleeves whenever working in or harvesting your blackberries.

If you prune them yearly, most upright blackberries can stand on their own without support, though many homesteaders prefer to grow them on trellises or between wire fencing simply as a matter of convenience. You must tie up trailing blackberries, such as loganberries, to a sturdy fence or trellis.

Cut canes back to ground level every fall after harvest is complete.

Insect pests and disease

Most insect pests and diseases can be kept to a minimum by practicing good cultivating sanitation such as cutting out old canes on a yearly basis. Choosing disease resistant cultivars and purchasing certified nematode-free plants is also vital.

Insect pests that can attack blackberries include aphids, leafhoppers, mites, fruit worms, scales, leaf rollers, various beetles, cane borers and white grubs.

Treat aphids and leafhoppers with carbaryl, leaf rollers with carbaryl or rotenone. Ladybugs and praying mantises will help to keep aphids under control.

A good way to deter Japanese and raspberry beetles is to plant garlic among your blackberry canes.

Scale insects should be treated with malathion in early spring. Dormant oil sprays may also be somewhat effective against scale.

Beetles can be controlled with the use of rotenone applied several times throughout the growing season, beginning with the appearance of buds.

The diseases that attack blackberries are verticillium wilt of roots, cane gall, anthracnose, and orange rust. Verticillium wilt produces persistent wilting of leaves with eventually die-back. When verticillium wilt is caught early it is possible to cut the canes back to the ground, burning the diseased canes. If verticillium wilt persists, it may be necessary to dig up and destroy the affected plants. Plant resistant varieties in their place.

Anthracnose, a disease common in cool, wet areas, first presents as brown spots on leaves and stems. Leaves eventually fall off. Affected plants must be destroyed, however, avoid doing so when plants are wet as that will only serve to spread the disease to other plants. Plants can also be sprayed with copper hydroxide just before flowering.

Harvesting

Harvest blackberries when they're a

deep shade of black. Don't delay in getting the berries off the plants when they are ripe or the birds will beat you to it. You can purchase netting to put over the canes to help protect your crop from the birds if they are a particular problem in your area.

Pick blackberries when they are at the peak of perfection; glistening, plump, black jewels.

Freeze them immediately or refrigerate them for fresh use. Blackberries make wonderful jam and jelly as well as pies and cobblers.

Blueberries

There are several different types of blueberries available for the homesteader to choose from. Blueberries need a moisture-retentive, acid soil. It is useless to try to grow blueberries in alkaline soil without first amending the soil. Blueberries do best in open, sunny locations, however, they can be grown in partial shade.

Blueberries can also be grown in containers very successfully on the patio or porch provided they get enough sunlight.

Do a soil pH test before planting your blueberries to be sure the soil will support them.

Planting guidelines

Plant blueberry bushes in the fall or early spring when soil can be easily worked. Dig holes 3 to 4 feet apart and set the young plants in them. Cover the rootballs about 1 inch deeper they were in the pots. Water plants in well, and cover with a thick layer of straw and well composted leaf mulch.

Maintenance

Fertilize every spring, about a month before new growth starts. Early each summer mulch with well composted manure and leaf compost, composted pine needles, or peat. Soil additives containing iron chelates help acidify the soil. Use as needed around your plants.

Blueberries need no pruning for the first three years after planting. After that, prune bushes each winter. Fruit is borne on the previous year's wood, so you should selectively prune blueberry bushes by targeting the 2 to 4 of the oldest shoots from each bush.

Insect pests and diseases

Blueberries rarely suffer large infestations of insects. Spraying blueberries as a preventative measure is usually not needed. However, if any insects do become bothersome, hand pick them if possible, use an insecticidal soap, or use an appropriate pesticide.

Blueberry plants suffer from chlorosis, which can be caused by too little acidity in the soil. Yellow mottling occurs on the leaves and plant growth is stunted. Few, if any, berries are produced. To treat this condition, add acidifying agents to the soil.

Harvesting blueberries

Blueberry rakes are sold at many garden centers and nurseries, or they can be ordered online or through gardening catalogs. Handpicking is useful if there are only a few bushes because it gives you an opportunity to closely observe your plants for signs of insect damage or disease. It also allows you to selectively pick your fruit, leaving more unripe berries on the bushes to mature.

Blueberries can be frozen or canned, and can be dehydrated successfully for winter use. Blueberry jam is delicious! If storing blueberries for a short time for fresh use, keep them in a covered container in the refrigerator and use them within a couple of days.

Currants

Currant bushes are highly ornamental, take up a relatively small growing area, and provide plenty of fruit for eating fresh or making into preserves and pies.

Red currants are the most popular currant available for today's homesteader, although both black and white currants do have limited availability.

Almost any water-retaining but well-drained soil will work for growing currants.

Currants are woody perennial bushes that will thrive in either full sun or partial shade.

Planting guidelines

Before planting, dig the bed over and work in good layer of well-composted manure.

A soil test is helpful because currants are especially susceptible to potash deficiency.

Plant in early spring or autumn. Dig holes slightly larger and deeper then the size of the pots, about 5 feet apart. Mix some well-rotted compost in with the dirt at the bottom of each hole. Plant the bushes at the same depth as they were in the pots, adding more compost to the hole before putting in bushes, if needed. Water bushes well.

Immediately after planting a one year-old bush, cut each ranch back to four buds from the main stem.

Maintenance

Water only during prolonged dry spells. Remove any suckers that develop from the main stem or the roots.

Currants are self-pollinating, bearing fruit at the base of one-year old wood, and on spurs of two and three-year old wood.

In late winter or early spring, feed with a generous portion of composted manure and mulch. Mulching helps control weeds and retains needed soil moisture. Don't cultivate with a hoe as this can damage the shallow roots.

Currant bushes are pruned often. Every summer, beginning when they reach two years old start pruning when the new growth starts to turn brown. Cut the laterals back to three or five leaves, just above a leaf joint.

In the second winter, cut out any branches that ruin the overall bush shape. Be sure to cut these branches flush with the stem to prevent rot from taking hold. Prune laterals back to one bud from their bases to encourage spur formation.

During the third and fourth years, leave some laterals on the bush to grow into branches so that the established bush has 8 to 10 main branches on a 6 to

9 inch leg. Cut other laterals back to spurs on a yearly basis.

During the third winter, cut out one half of the bush's new growth. In the fourth winter, remove about one quarter of the new growth. Prune back only about 1 inch annually thereafter.

Keep the center of the bush fairly open. As the oldest branches become less productive, Replace them with stronger, new shoots.

Insect pests and diseases

Red currant aphids cause leaves to curl or blister, often producing a reddish tinge. Shoot tips my be distorted. Control red currant aphids by applying a dormant oil spray in midwinter to kill eggs. Carbaryl spray will also help to control these pests.

Two-spotted mites cause leaf distortion and color changes. Tiny silver dots can be seen on the leaves. Use an appropriate pesticide that is labeled to treat this pest.

The tiny green currant worms can defoliate the bushes. Once you notice the leaves are being chewed, search for the worms and handpick them. Shaking the bush and raking up any worms that fall to the ground is a good method of control as well. As a last resort, apply an even dusting of rotenone.

The currant borer does its damage on the inside of the cane. The borer larvae feeds on the pith, then pupates and emerges in the spring as a moth. A reliable sign of borer damage is when the tips of canes wilt and drop their leaves. Cut the cane back, then burn the infested wood.

Wormy fruit is caused by the currant fruit fly. Infested berries will have a discolored area at the point the egg was deposited by the female fly. The damaged fruit usually falls to the soil, then the maggot burrows into the soil, and emerges as a fly in spring. Sometimes the infected fruit stays on the bush, so be careful to remove all fallen or damaged fruit.

Nectria canker is a fungus disease that causes the shoots or large branches to die back. Bright red spots then appear on the dead wood. If signs of Nectria appear on your currant bushes, you need to prune out and destroy all dead shoots to a point about 4 inches below any obvious sign of diseased tissue. Spray with Bordeaux mixture or lime sulfur solution after pruning. Feed, mulch, and water properly to improve vigor of plants.

Powdery mildew is a fungus infection that causes white powdery masses on the bushes. The masses eventually turn brown and shoots become distorted. Remove affected shoots in autumn. Spray with an anti-fungal that is labeled for use on powdery mildew after fruiting is over.

Birds are also a major problem with currants. Covering your bushes with bird netting will help discourage them from feeding on your currants.

Harvesting currants

It is best to pick currants as soon as they are fully ripe. To harvest, pick the entire clusters as the fruit is very delicate. These clusters can be frozen whole for later use if desired.

Red currants make great tasting jams and jellies and are also used in making wine. White currants are sweeter and more delicately flavored. Black currants are also used in making jams, jellies and pies.

Gooseberries

Gooseberries are shrubby bush fruits that are very ornamental with their long arching branches, colorful spring flowers and abundant berries. Gooseberries do well on almost any soil of average fertility, although they prefer more northern climates. Protect plants from late spring frosts.

Gooseberries will grow on soil that has poor drainage and will do well in either full sun or partial shade. They grow about 3 feet tall and up to 6 feet tall.

Gooseberries are self-fertile but you will probably want more than one bush to be sure of an ample crop.

Jostaberries are a cross between black currants and gooseberries that are available for a change of pace as well.

Planting guidelines

Double dig soil and work in plenty of ell composted manure . Dig holes, spaced about 5 feet apart, slightly larger and deeper than size of pots the plants were purchased in. back-fill holes with soil and water in well. Spread a generous layer of mulch around plants to help soil retain moisture and aid in keeping weeds under control.

Maintenance

Spread a new layer of compost and mulch on the soil around bushes every spring. Remove any suckers that develop at the base of bushes. Prune to maintain vigorous two, three, or four-year old wood. Prune oldest canes back to ground level.

A gooseberry bush is usually grown on a short trunk or leg of about 6 inches, from which the bush is continually renewed with new shoots arising at or near ground level. Allow stems to grow for 4–5 years, then selectively remove oldest stems to make room for new shoots. Snap off any branches that form along or below the leg.

Gooseberry thorns make harvest tedious. Selective pruning helps open up the middle of the bush to make pruning easier. Wear gloves and long sleeves when picking gooseberries to protect your skin from painful pricks. Gooseberries may successfully be grown in pots with proper attention.

Because some birds like to feed on the new buds, you may want to protect your gooseberry bushes with bird netting in late winter.

Remember, do not cultivate gooseberries with a hoe due to the shallow root system.

Insect pests and diseases

The currant aphid overwinters in the egg stage on plant stems. Eggs hatch in early spring, and the insects feed by sucking out the plant juices, resulting in stunted and distorted new growth. As leaves continue to develop, they will be crinkled, with downturned edges. As the aphids feed, they excrete excess sugar and water in small droplets called honeydew. Ants love to feed on this, and a black fungus, called sooty mold often grows on this honeydew. These aphids are small, green insects that are found in colonies. Other aphid species also occasionally feed on currants and gooseberries.

Aphids can be controlled by the use of natural insect predators such as ladybugs, and small parasitic wasps. Insecticidal soaps are effective in some instances, however, when infestations are more severe, dormant oil sprays can be effective.

Currant fruit fly occasionally attacks gooseberries. Infested berries will have a discolored area at the point the egg was deposited by the female fly. The damaged fruit usually falls to the soil, then the maggot burrows into the soil, and emerges as a fly in spring. Sometimes the infected fruit stays on the bush, so be careful to remove all fallen or damaged fruit.

Imported currantworm can defoliate bushes in a matter of days. Use rotenone to control this pest. This worm is the most serious insect pest of gooseberries. Foliage is consumed by the small, spotted, caterpillar-like larvae. The adults are sawflies about the size of a housefly. Two generations hatch each year, causing damage in the spring and again in late summer. Diazinon or malathion will control this insect. Start looking for damage shortly after the leaves have fully expanded. The second generation usually is less severe than the first and does not require treatment.

Gooseberries are hosts for white pine blister rust, which causes few problems for gooseberries, but is deadly for many pines. Gooseberries are banned in some regions because of this problem. Contact your local extension office to see what the regulations are for growing gooseberries in your area.

Powdery mildew can become a serious problem with gooseberries. This fungal disease appears on the young leaves, shoots and fruit as a whitish powdery mold, which later turns to reddish brown. The most successful treatment for powdery mildew is spraying lime-sulfur at the rate of 3 gallons to 100 gallons of water. The 1st application should be made when the leaf buds are opening, to be followed by applications at intervals of 10 to 20 days, until 3 to 5 applications have been made. In severe cases the tips of the diseased canes should be cut out and burned. Benomyl spray applied before the bushes flower and again after harvest will also help control this disease. There are also mildew resistant varieties that can be purchased.

Anthracnose can be found on currants

and gooseberries as numerous, small, brownish spots that are thickly scattered over the upper surface of the leaves. The leaves then turn yellow and fall off. With anthracnose, the bushes are often stripped of foliage completely before the fruit has had a chance to ripen.

Leaf spot is similar to anthracnose in its appearance and behavior. the treatment is the same for both diseases. Apply an antifungal that is labeled for use against these leaf spotting diseases. Read the label for use instructions and precautions.

Harvesting gooseberries

Gooseberries ripen slowly, usually over a four to six week period. Wear gloves to make harvesting berries safer and easier. Fresh gooseberries can be stored in the refrigerator for a few days until ready to use. Gooseberries can be canned or frozen for winter use. Gooseberries make wonderful preserves and pies.

Raspberries

Similar to blackberries, raspberries are cane fruits. They come in red, black, or yellow varieties and can be purchased in both thornless and thorned cultivars. Red raspberries are, perhaps, the best known and most widely available. Black raspberries are available through nurseries and greenhouses and mail order gardening catalogs as well as growing wild in many regions. Yellow raspberries are the sweetest and mildest of the raspberries, however, the yield per plant is somewhat lower than with other cultivars.

Planting guidelines

Raspberries prefer well-drained soil and will thrive in full sun or partial shade. Sandy soil will need to have plenty of organic matter incorporated into it before it will sustain raspberries. Raspberries need a plentiful supply of moisture throughout the growing season.

Although early autumn is the best time to plant raspberries, they can be successfully planted in the spring or summer. You should double dig and rototill the prospective raspberry patch, making sure you pick out all rocks and roots.

Homesteaders often plant their raspberries along wire supports that are stretched horizontally between vertical stakes. Each cane is then tied to the horizontal wire support. Dig a shallow trench about 1 foot wide and just deeper than the pot depth. A generous portion of well composted manure should be worked into the bottom of the trench. As you set the canes in position, spread out the roots evenly and trim off any damaged parts. Replace the soil in the trench, holding each cane erect in turn as the soil is placed over its roots and made firm. Do not cover the roots with more than 3 inches of soil. Water well and place a thick layer of mulch and composted manure around each plant.

Immediately after planting cut canes back to within 6 inches of the ground.

Maintenance

In Late winter, mulch the bed with a good layer of well composted manure. Then, in spring, as soon as the growth buds on the raspberry canes start to appear, cut the canes back to a visibly live bud about 10in above soil level. The idea of this is to leave just sufficient top growth to keep the roots active. In the second summer, when the fruit has been picked, cut down all the fruited canes down to soil level. New canes will rapidly replace these. You can dig up new young canes that start growing between the rows and transplant them into the bramble-line Any canes or suckers pruned should be immediately burned to help prevent the spread of disease.

All new canes should be tied in to the horizontal wires individually as they grow.

In the following February the canes should be tipped, making the cuts to growth buds from 4 to 6 inches from the tips. This will stimulate better growth lower down where the berries are less liable to suffer wind damage.

Autumn-fruiting varieties should have canes cut out in February and the new growths will then fruit the same year.

Routine spraying with a lime-sulfur dormant spray just before buds open in the spring will help control fungal diseases such as anthracnose, cane and spur blight, mites and scales.

Insect pests and diseases

Insect pests that can attack raspberries include aphids, leafhoppers, mites, fruit worms, scales, leaf rollers, various beetles, cane borers and white grubs.

Treat aphids and leafhoppers with carbaryl, leaf rollers with carbaryl or rotenone. Ladybugs and praying mantises will help to keep aphids under control.

A good way to deter Japanese and raspberry beetles is to plant garlic among your blackberry canes.

Scale insects should be treated with malathion in early spring. Dormant sprays may also be somewhat effective against scale.

Beetles can be controlled with the use

of rotenone applied several times throughout the growing season, beginning with the appearance of buds.

Spider mites are a common insect pest of raspberries. These mites are tiny sucking insects that are found on the underside of the leaves. The leaf damage from these mites appears as small, light colored spots. There may also be leaf distortion.

Good sanitation practices will help to keep insect pests under control. These practices include such things as keeping weed populations down by mowing and weeding. Remove the fruit-bearing canes at the end of every harvest and burn the old canes that have removed.

The diseases that attack raspberries are verticillium wilt of roots, cane gall, anthracnose, and orange rust. Verticillium wilt produces persistent wilting of leaves with eventually die-back. When verticillium wilt is caught early it is possible to cut the canes back to the ground, burning the diseased canes. If verticillium wilt persists, it may be necessary to dig up and destroy the affected plants. Plant resistant varieties in their place.

Anthracnose, a disease common in cool, wet areas, first presents as brown spots on leaves and stems. Leaves eventually fall off. Affected plants must be destroyed, however, avoid doing so when plants are wet as that will only serve to spread the disease to other plants. Plants can also be sprayed with copper hydroxide just before flowering.

Some of the most serious diseases of raspberries are mosaic diseases. These virus diseases show up as distortion of the leaves, yellow-green mottling, loss of production and fruit quality. The risk of mosaic viruses can be reduced by purchasing certified disease free plants from a nursery or garden supply center.

Strawberries

Strawberry plants make an excellent addition to the home garden. These attractive plants can be grown as a groundcover or landscape ornamental, in hanging baskets, strawberry pots, or in a traditional bed.

Choose vigorous, virus-free strawberry plants from an established nursery or garden catalog.

Three types of strawberries are readily available to the homestead gardener.

June bearing strawberries produce one large crop in late spring or early summer. June bearers yield the highest crop per season and the entire crop will ripen and be ready for harvest over a three week period.

Ever bearing strawberries produce one smaller crop of berries in the spring and another smaller crop in the autumn. Day neutral strawberries usually produce fruit all throughout the growing season.

Planting guidelines

Strawberries need full sun to produce the biggest, most flavorful crop, but they will also grow and produce a somewhat smaller crop in partial shade. Strawberries will not tolerate either drought or standing water. Make sure that the site you select is well-drained. Work a sufficient amount of organic material into the soil If it tends to be on either on the sandy side or contain a lot of clay.

The site you select should be free from weeds, insects and soil-borne diseases. Weeds tend to compete with strawberry plants for moisture and nutrients and can cause a marked reduction in production.

Strawberries are susceptible to Verticillium wilt, and should not be planted where eggplant, peppers, potatoes or tomatoes, have grown in the past three years since these vegetables often carry the disease. If you must plant your strawberries anywhere that these vegetables have grown in the past three years, you should choose disease resistant strawberry cultivars.

Double dig the bed two to three weeks before planting in early spring or fall. Work a 5 gallon bucket of well-composted manure into the soil for each 5 square feet of garden space. Just before planting weed the site and rake the bed to break up clods and level the soil.

Dig a trench that is 1 or 2 inches deeper than the roots of your plants. Form a slight mound of soil along the bottom of the trench. Dig trenches about 30 inches apart. Set plants 18 inches apart in each trench with roots gently spread out over the mound. Position the upper part of the roots, called the crown, level with the soil surface. If the crown bud is buried beneath the soil surface it can rot. If it is planted too shallow, the roots can dry out and die. Fill the trench with soil and firm around each plant carefully. Water plants in well.

Maintenance

Water plants regularly in dry weather, especially during the first few weeks after planting. Water stress will seriously stunt the growth of new strawberry plants or kill the plants entirely.

In autumn cut off any runners that have grown over the summer. Cultivate to remove weeds in early spring, and apply an even layer of well-composted manure, topped with a layer of straw. Plastic sheeting or layers of newspaper can also be used as a mulch to retain soil moisture and help protect the developing fruits from prolonged contact with the soil.

With Junebearing strawberries, once the harvest is over, remove plastic sheeting if any was used. Rake the straw or newspaper well up onto the plants. Set fire to the straw or paper and allow the strawberry plants leaves to burn off. This will in no way harm the plants but will burn the old leaves, killing any disease organisms and destroying insect pests that may be present. New leaves will soon grow to replace the old.

If you prefer not to burn the leaves, you can cut the plants down with shears to about 3 inches above the crown. Pull of any unwanted runners and old or diseased leaves. Rake off all litter and burn or otherwise discard it.

Everbearing and day neutral strawberries should not be cut back or burned over. Remove only diseased or dead leaves from these plants.

With all strawberries, a thick mulch of dead leaves helps protect the crowns from heavy winter freezes in more extreme climates.

Renew strawberry plants every three years. Older plants are more susceptible to diseases and insects, and their production declines with age as well. New replacements can be grown from existing plants.

In early summer choose strong, healthy parent plants that have been producing well. Choose three or four strong runners from each plant and extend these runners out from the plant. Fill 3 inch pots with good loam and sink the pot into the ground even with the soil surface just beneath a healthy tuft of leaves along the runner, closest to the parent plant. This tuft of leaves is the embryo plant. Using a small U-shaped piece of galvanized wire, secure the runner to the soil placing the wire on the side of the embryo closest to the parent plant. Pinch off any extra growth beyond the embryo plant along the runner. Do not sever the runner and embryo from the parent plant. Keep the soil moist around the rooting embryo plant. These baby plants will be ready for transplant in about 4 to 6 weeks.

In late summer sever the runner connecting the new plant to its parent plant. Continue to keep the new plant evenly watered. One week later carefully lift out the pot containing the new plant. Transplant the new plant into it's permanent position in the strawberry bed.

Insect pests and diseases

You may find slugs and snails eating your ripening strawberries before you have a chance to pick them. Setting out half buried bottles of beer to trap slugs can help reduce the number of slugs and snails in your garden. Occasionally mice and birds will become a problem. Keep your strawberry patch picked daily to discourage these pests.

Aphids can cause leaves to become distorted and turn yellow. Insecticidal soaps and carbaryl will help keep these insects under control.

Strawberry leaf beetles feed on leaves of strawberry plants, while their larvae damage the plants' roots.

Powdery mildew is a fungal disease that causes the leaves of the plants to turn purple and curl upward, exposing the underside of the leaves. This fungus can be controlled by spraying with a 1 ½ % lime sulfur solution just before plants blossom and again every 2 weeks until 2 weeks before harvesting berries.

Harvesting Strawberries

Birds and rodents love strawberries, so be sure to pick over your plants daily during the harvest season, usually beginning in June. It is especially important to keep your strawberry plants picked during rainy spells to prevent fruits from rotting.

Sort freshly picked berries. Use fully ripe fruit first, leaving under-ripe fruit out to ripen for another day or so. Strawberries can be stored in the refrigerator for only a day or so before using.

Strawberries can be dehydrated, or frozen for best quality. Strawberries may also be canned, however the quality is not as good as the dried or frozen product.

Strawberries can be used alone or in combination with other fruits to make wonderful jams, jellies, preserves, pastries and table syrups.

Grapes

Whether grown for fresh eating, to dry for raisins, preserve in jellies, juice, or wine, grapes will add healthful variety to the homestead pantry. Entwined along a covered trellis, called an arbor, grapevines can also lend a shady respite to any homestead landscape design.

There are many varieties of grapes that fit into one of three types.

The European grapes such as Thompson Seedless, are characterized by skins that stick to the flesh of the fruit. These grapes varieties prefer warmer, drier climates.

The native American types, such as Concord, have skins that slip free from the flesh of the fruits. These grapes can be grown in a wider range of climates than can the European types, however, they do best in cooler climates.

There are also the muscadine types, such as Southland, which are not noted for their hardiness. These grapes prefer a warmer, more humid environment.

On a grapevine, the main stem is called the trunk. The trunk sends off laterals that develop into arms, spurs, and canes. It is the canes that bear the fruit. One grapevine needn't be limited to only one trunk. One vine can have two or even three trunks, although limiting the number of trunks to only one will help to conserve space so that multiple varieties can be grown in a smaller area.

Grape vines can be trained to grow on a fence, trellis, or arbor. European types can grown as very small trees that are supported by stakes.

Planting guidelines

When selecting your grape planting site, remember, it is a permanent bed, so choose the site with care and forethought. Good drainage is essential for grape cultivation and production. Full sunlight is a requirement for fruit with a high enough sugar content for juice or wine, or fresh eating.

Grapevines can be planted either in early spring or in the fall. Dig a hole slightly deeper and wider than the rootball of the plant. Plant the young vines with the top bud level with or slightly below the ground surface. If planting a large number of grapevines, plant the vines about 6 feet apart in rows that are spaced about 8 feet apart. If planting vines to grow up a covered arbor, plant one grape variety on either end of the arbor.

After planting, firm the soil well around the base of each plant, and water the plants in well. Spread several inches of well-composted manure mixed with straw as a mulch around the plants.

Muscadines can be male, female or self-fertile. You will need 1 male or self-fertile vine for every 6 to 10 female vines. Of course, if growing only self-fertile varieties, you will not need a pollinator. Plant muscadines in an area that gets at least a half day of sunshine. Allow 20 running feet of space per vine along a wire trellis, fence or arbor.

Maintenance

Grapes need a lot of moisture and are fairly heavy feeders. Replenish the manure and straw mulch layer every spring.

Training young grapevines to an arbor, fence, or trellis is easy, however it can take up to four years to get them properly trained. After four years, the vines are considered mature and all that is needed is pruning to improve fruiting.

Allow new vines to grow for one season without cutting them back at all. This will encourage the plants to develop healthy root systems. Keep the plants watered well during the first year.

Grapes are both ornamental and useful when trained to climb up a trellis or arbor. During the first growing season, remove all but the strongest single shoot. Tie this shoot loosely to a stake. This will become the vine's trunk. During the first dormant season after planting, select a few of the strongest canes to keep and prune out all of the others. Train these strongest canes evenly up the trellis or arbor. If the canes are several feet long you may want to tie them to the trellis in several places along the length of the canes. Allow these well-spaced canes and their spurs to grow, continuing to train them to the support system as they grow.

On a yearly basis, keep removing old fruited canes each winter so that the new canes of the previous summer can grow to fruit during the next growing season. Cut new canes back to 6 to 10 buds, depending on the vigor of each cane. Shoots that appear at the base of new canes should be cut back to only two buds to form the spurs for the next season's growth.

During the second summer, prune out all but the strongest two shoots, or the strong leader shoot and one strong lateral. Tie these shoots along the trellis or arbor so that they will grow in opposite directions. Let these shoots grow to about 18 inches, then, pinch back the tips.

After the second growing season, prune out only the excess shoots. Aim to keep form and spacing even as you select the canes to keep and then gently tie them onto the trellis or arbor.

In the years that follow, keep removing old fruited canes every winter to keep the new canes of the previous summer healthy and strong for productive fruiting over the coming growing season. Cut the new canes back to 6 to 10 buds apiece, depending upon the vigor of the new canes. Repeat this pruning process every winter.

You can improve the yield of your grape vines by thinning the flowers and fruits during the summer months. Grapes can be thinned by using one of three methods. Some of the flower clusters can be removed from the vines in the spring to ensure that those remaining will develop into larger, more compact bunches. You can also wait for the fruit to set before thinning out the smaller clusters, thereby ensuring that the remaining clusters will contain larger, more flavorful grapes. Grapes can also be thinned by targeting individual grapes within a cluster to thin. This is a tedious endeavor, however, it will produce excellent bunches of larger, more flavorful grapes. Shortly after the fruit has set, use scissors to remove the lower tip of each bunch and one or more weaker stems within each cluster. Also thin out any grapes or clusters that show signs of disease or insect damage.

Bear in mind that although the large leaves of the grape vine help shield the developing fruit from the hot midsummer sun, too much shade later in the growing season can inhibit ripening. Therefore, in late summer, pinch foliage-bearing side shoots back to only a single leaf.

Prune muscadine grapes by cutting back all new growth to three or four bud spurs on the long arms. Clusters of grapes will appear on the third to sixth nodes of that season's new growth. Prune after the first severe freeze, but not later than the beginning of the new year.

Do not use insecticides during the time of blossom on muscadine or you risk killing the honey bees that are vital to the pollination of many of these varieties of grapes. Carbaryl is an excellent insecticide for use on grapes that should be used before and after the bloom.

Sugar content of muscadines increases as the grapes begin to ripen and turn their characteristic color. Delay the harvest a little while and you will have extra-sweet grapes that are superb for wine or juice.

Insect pests and diseases

Grape berry moth larvae feed on the inside of developing fruits. They often ensnare several grapes together in a silky webbing. Carbaryl, used just after the bloom, will control this pest. Be sure to follow up with another application 10 to 14 days after initial use, or immediately after bloom is over. Repeat again in mid or late summer for continuous control.

Leaf rollers and leaf tiers are moth larvae that feed on foliage of new shoots, chewing and scarring them badly. These larvae may also attack fruits. Leaves are often rolled up with a sticky webbing on the inner surface of the leaves. Control these insects with carbaryl used in early summer. Follow up in 14 days with a second application.

Japanese beetles and rose chafers also attack grapes, feeding on the foliage to the point of skeletonizing the leaves. Carbaryl offers some control. Use the same as for use against grape berry moths. Japanese beetle traps set about 100 feet away from your orchard and grape arbor can help lure beetles away from your fruits and into the trap for disposal.

Black rot is a fungal disease that attacks the leaves of grapes in early summer and can also appear later on half-grown fruits. Entire grapes will shrivel to dry, black, raisin-like fruits. Infection with black rot usually occurs during the bloom period. Captan or another properly labeled fungicide applied just prior to the bloom, again after bloom, and then every 14 days thereafter until early fall will help keep this disease under control.

Dead arm is a very destructive fugal disease that first attacks the leaves and canes. Later it infects young shoots. Dead arm phase results from infection through wounds caused by pruning or from winterkill. Infected vines should be removed at ground level during the regular winter pruning. Spray with captan when new shoots are about 1 inch long and again when they reach 5 inches long. Remove any infected debris and burn it.

Downy mildew first appears a small yellow spots on the tops of leaves. Later the downy white mats that give the fungal disease its name appear on the lower leaf surfaces. Complete defoliation may occur. When fruits are attacked they turn to an off-color, and harden. Use the same fungal control measures as used for black rot. Some varieties are resistant to downy mildew.

Powdery mildew occurs on foliage in early summer and fall, producing powdery growth on the leaves as well as fruits of grapes. Planting grapes where they will get plenty of sunshine, proper pruning, and good drainage will help prevent this fungal disease from taking hold in your vineyard or arbor. Sulfur or other properly labeled antifungal spray can be used to treat vines for powdery mildew.

Harvesting grapes

The best way to see if your grapes are mature enough to harvest is to taste them. If they are sweet and flavorful, they are ready to pick. Color is also an important test for ripeness. Green varieties should turn whitish or yellowish green. Black and red varieties will take on an added depth of color. Concord grapes will have a deep purple color with a slight whitish blush to the skins.

Once picked, grapes will not ripen further. Be sure your grapes are fully ripe before harvesting.

Once fully mature, pick entire bunches with scissors or a sharp knife. Discard any grapes that are overripe, withered, hard, or show any signs of insect or disease infestation.

Grapes that are intended for drying should be left on the vine somewhat longer than wine, juice, or table grapes in order to increase their sugar content.

Grapes can be stored for several weeks at a temperature of about 32 degrees F and a humidity of about 90 percent.

CHAPTER FIVE
In the Orchard

UNLESS YOU'VE EVER GONE out and picked a sweet, ripe cherry from one of your own trees and popped it, succulently fresh, into your mouth, you cannot begin to know the joys of fresh fruits from your own orchard.

Apples, apricots, cherries, nectarines, peaches, pears, and plums can all be grown on nearly all homesteads. Only the warmest of climates will not sustain apples, pears, or cherries, while peaches, plums, and nectarines can be grown in all but the coldest of climates.

The homesteader with several acres on which to plant the orchard will have the freedom to try several different types of fruit trees, both self fruitful trees as well as fruit trees that require a pollinator.

For the homesteader with only enough space for a small orchard, or even someone desiring to plant a few trees in an edible landscaping design, there are several self-pollinating fruit trees available. Peaches and nectarines are self- pollinating, as are apricots, damsons, pie cherries, and even a few varieties of apples. Apples and pears are now available on "family" trees, or trees onto which several different varieties of apple or pear branches have been grafted. These trees usually develop from three to five different varieties of apples or pears on one tree.

Fruit trees are generally available in three forms; standard, semi-dwarf, and dwarf types.

Standard trees are full sized trees that produce large crops of fruit, but take years to mature to fruiting size.

Semi-dwarf trees are smaller trees that fit into smaller areas easier and produce earlier, however, they also yield smaller crops than standard trees will produce.

Dwarf trees are the smallest fruit trees available, producing small but tasty crops of fresh fruits in the smallest amount of space. Earliest to produce, easier to prune and harvest fruits, dwarf trees are popular with homestead gardeners.

Do not expect dwarf varieties to be as hardy as semi-dwarf or standard varieties though. Standard trees are the best choice for the coldest extremes, while dwarf trees are often the best choice for warmer climates. Your local County Extension Office can provide you with information regarding which varieties of orchard fruits grow best in your area.

The availability of disease-resistant varieties of fruit trees may help make the choice of which varieties to grow a much simpler one. Less disease resistant varieties that require more spraying, especially in regards to apples and pears may not be as desirable, regardless of their flavor.

Insect control on fruit trees should begin with pre-bloom sprays, either dormant or delayed dormant, for pests such as scale insects, aphids and mites. Applications of a horticulture oil in combination with an appropriate insecticide will control these insects if applied during late winter or early spring before the buds open. Do not apply dormant oil if a heavy freeze is expected as damage to the tree could result.

Timing is vital because oil sprays need at least 10 to 12 hours to dry before temperatures drop below freezing. Oil sprays will also darken tree bark and buds, speed up spring bud development, and reduce the ability of buds to withstand colder temperatures. Therefore, it is best to delay dormant oil sprays until the buds start to break. Make certain that branches in the center of the tree receive adequate spray coverage. Base the amount of spray on the size of the tree, unless instructed otherwise. A poorly pruned tree may need up to double the amount of spray for proper coverage compared to a properly pruned tree.

Several all-purpose sprays made specifically for fruit trees are on the market for home use. These sprays contain an insecticide as well as a fungicide that will control most insects and diseases seen in the homestead orchard. As with any pesticide, read and follow all label directions and precautions.

Homesteader should also follow such cultural practices as proper sanitation to reduce the incidence of insect pests. Good sanitation includes promptly removing and destroying dead, diseased, insect infested, and damaged wood, leaves, and fruit. Any leaves, wood and fruit left on the ground will often provide pests with places to complete their development or to survive the winter.

Although adequate insect control on peaches usually requires spraying trees, these sprays need to be timed accurately to be effective.

Knowledge of the insect pests that affect fruit trees and their life cycles will aid in identification of, as well as the early diagnosis of, any developing pest problem and the accompanying ability to treat it quickly.

Apple Trees

Apple trees available commercially are either budded or grafted trees. The selected variety or top of the tree, referred to as the scion, is grafted or budded onto a rootstock of choice. The rootstock determines the relative size of the tree but it does not in any way affect the type or quality of the fruit that the tree will bear. With currently available commercial rootstocks, there is a wide range in tree size potential as well as some degree of resistance to certain root borne insects and disease problems.

Popular apple varieties

- Cortland is a late season apple. A cross between Ben Davis and McIntosh, Cortland is a bit larger, and has more attractive color. Cortland has a wonderful perfume, and a slightly acid, very good flavor. Cortland is, with McIntosh, the archetypical American apple. It is good for pies because it doesn't turn brown when it is sliced. Cortland is

a vigorous tree with high yields.
- Empire is a late season apple. A cross between Red Delicious and McIntosh, Empire is medium sized and sweet, but with the acidity, and the crisp flesh of the McIntosh.
- Fortune, also known as Laxton's Fortune is an early to mid season apple. It is sweet and aromatic with excellent flavor. Fortune is a crisp apple early on and becomes softer the longer it is left on the tree. The trees spur freely, and are remarkably resistant to most apple diseases with the exception of European Canker
- Freyburg is a mid to late season apple. Usually small to medium in size with dry skin that is light golden yellow and slightly russetted. Freyburg is an extraordinary flavored apple. The flesh is crisp, very juicy, and light yellow. It is sweet and moderately rich flavored with a distinct anise flavor. The anise component becomes stronger the longer the apple is left on the tree, and varies with seasonal climatic conditions. Freyburg stores well.
- Golden Delicious is a mid to late season apple. A tree ripened Golden

Delicious apple is juicy, sweet, honeyed, and of excellent flavor and texture. The tree is highly productive, bearing on spurs, laterals and tips. The fruits are medium sized, clear yellow, sometimes lightly russeted. The trees flower heavily and over an extended period, making it an excellent pollenizer for other apple varieties. The tree is a vigorous one. There is a 'spur' form of the tree, which is a little smaller, but just as productive.

- Granny Smith is a late season apple. This very late maturing, late keeping, large, green, dual purpose apple has hard, crisp and juicy flesh. The flavor is tart, becoming very sweet if left to ripen on the tree. Granny Smith apples are excellent keepers. The tree is very vigorous and crops heavily, but it is not recommended for areas with very short growing seasons. Granny Smith is also an excellent pollenizer for other varieties.
- Jonathan is a mid season apple. The medium sized fruits have thin, bright red-blushed skin that contrasts sharply with the crisp, juicy, yellowish white flesh. The flavor is deliciously sweet-tart and aromatic,

and the texture is crisp. Jonathan is an excellent dual purpose apples with fair keeping qualities. The trees are vigorous growing but at maturity are fairly small. Jonathan is more disease susceptible than some other varieties.

- McIntosh is a late mid season apple. It is a large, bright red apple with a heavy bloom to the fruit. McIntosh is the archetype of the American apple with its perfumed, crisp, juicy, snowy white flesh. The flavor is very good, being sub-acid and perfumed. McIntosh is a good keeper and a well loved, dual purpose apple. It is susceptible to apple scab.
- Red Delicious is a mid to late season apple. This medium sized, tough skinned, juicy, and somewhat coarse textured apple sets the standard for fresh eating apples. Red Delicious makes a good pollenizer for other varieties.
- Rome Beauty is a late maturing apple. It has skin colored red over a yellow ground color. It is a large, attractive cooking apple that could be eaten fresh. The trees are vigorous and quick to start fruiting, regular and reliable, however, they

are also late, and susceptible to apple scab and fire blight. Rome beauty makes a fairly decent keeping apple.
- Stayman Winesap is a late season apple. These fruits are medium to large, with a thick, tough skin that is yellow with dull red stripes. The flesh is fine-grained, tender, yet crisp, juicy, aromatic and sub-acid. The tree is vigorous and productive. The fruits are fairly good keepers.

Site selection and planting guidelines

Select a sunny location with rich, well-drained soil for planting your apple trees. Planning your apple orchard so that the trees will get early morning sun will help reduce incidence of powdery mildew and other fungal diseases.

Early spring is a good time to plant apple trees in the northern climes.

In the South, fall is the best time for planting apple trees, as fall planting allows the roots to become established when the following spring comes around, thereby giving the young trees a head start.

Plan your orchard by marking spots for planting holes, spacing them 6 to 8 feet apart for dwarf varieties, 12 to 16 feet apart for semi-dwarf varieties, and 18 to

25 feet apart for standard varieties.

Before digging the planting holes, remove weeds and grass to form a bare-soil circle for each transplant, about four feet in diameter. If planting bare-root stock, soak the sapling's roots in cool water for about 30 minutes before planting. If the roots look particularly dried out, you can extend the soaking period to about 24 hours.

Begin digging a hole approximately twice the diameter of the root system, and about a foot deeper, if planting bare-root stock. If planting potted trees, dig the hole about 2 inches wider and deeper than the size of the pot. Spread out the roots in the hole as much as possible and check the level of the "bud union." The bud union should sit about 2" above soil level. The bud union is where the scion, or cutting, meets the rootstock as a result of grafting. You don't want the bud union to be buried because you don't want to invite crown rot. You also don't want the scion taking root and overriding the job of the rootstock.

Water well as you fill the hole back in with soil. Mix well composted manure to the soil that you backfill with and tamp it down well after watering to remove air pockets. This is the perfect time to install a vole or a guard around the trunk of your apple trees. Flexible, drain pipe cut to about 1 foot in length is perfect. Slice the pipe up one side and place it over the young tree trunk, letting it protrude about 10" above ground level. Water well again after the transplanting is complete. Place a 3 inch layer of mulch around the base of the tree, being careful not to heap it too high directly around the trunk.

Maintenance

Dwarf type apple trees should be staked or supported for optimal tree growth. Staking also helps to support the tree and the fruit load in the early years when the young apple tree may be weakest. Ten foot wooden stakes, about 2 inches in diameter, are generally recommended with 2 feet driven into the ground and the remaining 8 feet above the ground where the tree's main trunk is loosely tied, approximately every 2 feet, along the length. One inch aluminum conduit also works well as a support for dwarf apple trees.

In pruning apple trees the goal should be to try to give them an optimal shape and structure. Focus your efforts on the leader and on establishing good scaffold branches. A prime objective in pruning apple trees is to ensure good aeration. If air can circulate freely through all the branches, there will be less likelihood of a problem with powdery mildew or other fungal disease. Pruning also restricts the vertical growth of apple trees, making tree care and harvesting fruit easier.

In addition to pruning, the branches are trained in a process called, "spreading" to form angles that will help the branches to radiate out from the trunk, while maintaining sufficient strength to bear heavy fruit loads. Keep all this in mind when selectively pruning your apple trees. Use your own hand as an example. Crotches like the thumb are good, while crotches like the fingers are bad.

Thinning is also an important part of apple tree care. Too much fruit can be detrimental to the tree's health. Thinning fruits promotes larger fruit size, improves next year's blooming and reduces the likelihood of limbs snapping off under the pressure of a heavy fruit load. Thinning works on two levels: the blossom level and the branch level.

Apple blossoms form a flower cluster consisting of five or six potential fruits. You'll want to thin these fruiting clusters down to one or two fruits each, once the tiny apples have reached about the size of a marble. At the branch level, remove enough fruit so that the remaining apples are spaced about 4–6 inches apart.

Insect pests and diseases

Codling moth is the most serious insect pest of apples. Thinning fruit, trapping the insects, and appropriately timing insecticide applications can control this insect in home orchards.

The larva of the codling moth is the worm responsible for "wormy apples." Wherever these fruit trees are grown, codling moth is a problem to some degree, and usually will need some form of treatment to protect the fruit.

Codling moths have two and maybe sometimes even three generations per year. Adult moths first emerge during mid-spring, usually around the period of full bloom. This flight continues for about six or seven weeks, peaking a couple of weeks after the first moths become active.

These moths lay eggs on the leaves of the trees. The larvae first feed on the leaves, later moving on to the developing fruit. After feeding within the fruit for a length of time, the full-grown larvae migrate out and crawl down the trunk of the tree to pupate under bark flaps or in other protected spots.

The second, and most important, generation occurs in early summer. Adults emerge in early July, laying eggs directly on the fruit surfaces. The young caterpillars tunnel into the fruit shortly after the eggs hatch. Most of the fruit damage occurs at this time. Again, when full grown, these caterpillars leave the fruit to seek other protected areas to complete their development.

Insecticides are a common technique used to control codling moth and should be applied shortly after petal fall. Many people continue these treatments at 10- to 14-day intervals throughout much of the summer. Permethrin and carbaryl (Sevin) are the most common homeowner treatments for codling moth. Remember to never apply insecticides during periods of bloom. Treatments at this time will kill beneficial pollinators, such as honeybees.

Insecticide use can be limited and treatments better timed by using pheromone traps. These specialized traps are baited with the sex attractant of the female codling moth. The traps capture male moths and give an estimation of when mating and egg-laying take place. The first insecticide application often is optimal about three weeks after the first consistent captures of male moths. Pheromone traps that only collect males cannot control codling moth, although they can aid in pinpointing the proper timing for insecticide use

There are several non-chemical approaches that also help in the control of the codling moth.

Thinning fruits can limit codling moth infestations. Thinning also helps provide better control of codling moth when sprays are used in conjunction with thinning. Pick up and dispose of any apples that show evidence of codling moth entry wounds.

A burlap or corrugated cardboard band around the trunk can cause many larvae to settle and pupate at these sites, where they can be regularly collected and destroyed.

Some control of adult codling moths is possible using homemade traps containing water and molasses mixtures, typically in about a 10:1 ratio. These water-molasses traps capture both sexes.

Apple maggot is a potentially serious pest of apples. The maggots tunnel into the fruit, creating characteristic brown tracks.

Apple maggots spend the winter in the pupal stage scattered around the base of apple trees. The adult flies emerge in midsummer and lay eggs on the developing fruit. There is only one generation per year.

Apple maggot flies can be trapped with yellow sticky boards or red sticky balls that can be purchased specifically for use against this insect. The use of several of these red sticky balls per tree can provide substantial control of this insect by killing adult females before they lay all their eggs. When used in conjunction with insecticide use, a good level of control can be managed against apple maggots.

Scale insects sometimes damage apple trees. Perhaps most damaging is the San Jose scale, a tiny, circular scale found on fruit as well as branches. Heavy infestations can cause death of infested branches and spotting of infested fruit. It can also contribute to a general decline of the tree.

Scale infestations are typically slow to develop. Regular treatments are rarely needed in backyard orchards. Treatments with horticultural oils during dormant periods in early spring can provide adequate control of these insects.

Several species of aphids can damage the new growth of apple trees, while the rosy apple aphid may injure developing fruit, causing it to become mottled and misshapen.

Aphids on fruit trees over-winter as eggs clustered near the dormant buds. After the eggs hatch in spring, the aphids feed on the emerging leaves, causing he leaves to curl slightly. Horticultural oils applied in late winter or early spring can prevent these problems.

Diazinon, malathion and insecticidal soaps also can help suppress aphid outbreaks after leaves emerge. However, aphid infestations typically are short-lived on apples. Aphids usually leave the tree by late spring, flying to alternate summer host plants. If aphids are not seriously damaging the majority of terminals or injuring fruit, there is little benefit from insecticides for aphid problems after leaves emerge. Beneficial insects such as lady bugs can also help keep aphid populations under control.

The woolly apple aphid affects apples in differently than other aphids do. This insect has long, thread-like strands of wax that cover its body.

Woolly apple aphids infest twigs, callous tissues around pruning cuts, and sometimes even the roots of apple trees. Sustained infestations of woolly apple aphid over several years cause cankers to develop that can girdle and kill parts of the plant. Insecticidal soaps and carbaryl are effective against the woolly apple aphid where it occurs on stems. There are no effective treatments for woolly apple aphid on roots.

Apple diseases such as scab, fire blight, cedar apple rust and powdery mildew can cause severe leaf loss, poor fruit quality and general tree decline. Severity of these diseases will vary with location and weather conditions. A fungicide spray program is often needed to protect susceptible trees and fruit. To minimize disease severity as well as the number of fungicide applications in those areas where disease is known to be a problem, you should choose disease resistant apple varieties when planting new trees.

Apple scab is the most widely distributed of these diseases Apple scab is caused by a fungus that infects the fruit and the foliage of apple trees under cool, humid conditions in the spring. Young, velvety brown lesions can be seen on the underside of leaves. With time, individual lesions may coalesce and infect both the upper and lower leaf surfaces. A severe infection of the leaves can cause premature defoliation, which reduces tree growth and yield.

Scab lesions on the fruit are brown and corky. As the fruit enlarges, it may grow unevenly, resulting in misshapen, cracked fruit. Fruit losses from apple scab can be severe on susceptible varieties.

Cedars act as an intermediate host for the disease known as cedar apple rust. The spores of cedar apple rust are wind-blown from cedars and related trees to susceptible apple trees. After infection, orange-brown lesions appear on the upper sides of the foliage or on fruit. On susceptible varieties, cedar apple rust can cause complete defoliation and loss of fruit quality.

Fire blight is a devastating bacterial disease that occurs sporadically in apple trees. This disease infects the blossoms, fruit, branches and leaves of susceptible apple trees. The infected tissue appears black, as if scorched by fire, and is often accompanied by clear or milky ooze. The "shepherd's crook" symptom, in which the shoot tips are bent over, is the most easily recognized evidence of this disease. Whole branches or even entire trees may be lost after a fire blight infection. Temperatures greater than 65 degrees F and high moisture favor fire blight infection.

Powdery mildew is caused by a fungus that infects the blossoms, fruit and leaves of susceptible apple trees. Whitish patches can be seen on the underside of foliage. Infected flower buds open five to eight days later than the healthy ones. Later, the developing fruit often exhibits russetting, which appears as brown, corky netting on the surface of the small apples.

Powdery mildew infection favors cool temperatures and high humidity.

Anthracnose is a fungal disease that sometimes occurs in cooler climates. Anthracnose forms cankers that are often called fiddle-string cankers because of the long, string-like fibers of the inner bark that are left exposed in the center of the canker. These cankers usually occur on smaller twigs and branches and expand for only a single season. The fungus produces spores in these cankers, which are washed and splashed to other branches and trees during rains or irrigation. Infection occurs most readily any time weather is cool and damp, especially in the fall.

Although some apple cultivars have varying degrees of resistance to some apple diseases, they are still susceptible to summer diseases, such as fly speck and sooty blotch.

Fly speck and sooty blotch occur together on the fruit surface under warm, humid weather conditions. Fly speck can be identified by distinct groups of tiny, shiny black spots on the fruit surfaces caused by the excretions of an insect. Sooty blotch appears as olive green to black smudges on the surface of the fruit.

Both of these diseases are superficial blemishes that can usually be removed from the surface of the apple with mild scrubbing. Although all apple varieties are susceptible to infection by both fungi, symptoms are most severe on yellow or light-colored varieties such as Golden Delicious or Grimes. To aid in prevention of these diseases, select an orchard site that always has full sunlight, good air circulation, and good drainage. Pruning trees annually to an open center for maximum air circulation will also aid in prevention of both of these diseases. Both diseases are most prevalent in the damp, low, shaded areas of the orchard. Any practice that opens up the trees to greater air movement and promotes faster drying greatly aids in disease control.

Finally, you may want to consider bagging the fruit when it reaches the size of a golf ball. Purchase a quantity of very small paper bags. Fasten one paper bag carefully onto each fruit, using a twist tie on the stem to fasten. This process will prevent any late-season insects from infecting the fruit and will also prevents cosmetic skin infections like sooty blotch and fly speck.

A week or two before harvest, remove the paper bags to allow the fruit to color in the sun. You will get fruit of the very finest quality that has not received pesticides of any kind for at least a two month period prior to harvest. Never attempt to use plastic baggies to achieve the same goal.

Plastic will retain moisture, causing rot to develop on the fruit.

Harvesting apples

Apples ripen at different times of the year from mid-summer through late fall. It is not always easy to determine when is the right time to harvest apples. If you know what to look for you can harvest apples at their peak. Apples that are picked at the right time keep longer in storage and have better flavor.

Pick apples before they are fully mature. You can gauge apple maturity several ways. First, look for changes in fruit skin color. When most apples mature, the background skin color changes from a leafy green to a lighter shade of green and eventually to a yellowish color. You can pick most apple cultivars when the first signs of yellowing appear on the fruits. For red-skinned fruit varieties, the flesh will change color from greenish to white when fruit maturity is reached, however, the greenish color of certain varieties of Red Delicious flesh may only disappear after several months of storage.

Fruit crispness, and loss of a starchy taste are two other signs to look for when determining ripeness for harvest. Soft grainy flesh means over-ripe apples. Likewise, hard flesh is a sign of under-ripe fruit. Perfectly mature flesh should be deliciously crisp and juicy.

You can also tell if your apples are ready for harvest by how easily the fruits detach from the tree. Because of this, when the apples on the tree are nearly mature, you may find a few sound apples that have dropped to the ground. These drops are still edible and are good forecasters for impending fruit maturity.

Mature apples separate easily from the spur or twig. Never shake the fruit from the tree. To pick an apple, grasp the fruit and lift up with a slight twisting motion. Avoid pulling down on apples when you pick them as this removes the fruiting spur along with the fruit. Spurs are very short branches that bear fruit every year. If you remove too many over time you'll reduce future apple production.

Sort through freshly picked fruit and remove any bruised and damaged fruit. Plan to use damaged fruit quickly because it is unfit for storage.

Store only sound apples in clean wooden or cardboard boxes that are ventilated to allow air to circulate. It is not necessary to line the boxes with paper or individually wrap the fruit.

An older but still working refrigerator makes a good fruit storage place. Ideally, storage temperature should be 30 to 32°F, however, such conditions are usually difficult to maintain on the homestead.

An unheated garage, shed, or basement may be satisfactory if temperatures below 30°F and above 45°F can be avoided.

An insulated box, storage cabinet, or dug-out underground room that can be ventilated at night for cooling also makes a good storage place for apples.

Maintain high humidity in storage by placing the fruit in unsealed or perforated plastic bags. Placing an open pan of water in the storage place will also increase the humidity.

Shriveling of Golden Delicious apples can be avoided by storing them in loosely tied plastic bags.

Store fruit immediately after it has been picked. Do not store apples with onions, potatoes, or other strong-smelling items because the fruit will absorb flavor from them.

Make it a habit to regularly inspect stored apples for mold, flesh breakdown, freezing, or excessive ripening. Apples held too long will become undesirable because the flesh will become soft and mealy and may have internal breakdown.

Storing apples with other fruits and root crops will cause the other crops to ripen and spoil quickly in storage due to gases given off by the apples. Apples should be separated from other stored crops for this reason.

Pears

There are three basic types of pears grown in the United States.

European or French pears include such popular varieties as Bartlett, Bosc, and D'Anjou pears. All of the European varieties are especially noted for their excellent fresh eating quality. However, the European pears generally have a higher susceptibility to fire blight. Oriental hybrids include such well-known varieties as Orient and Kieffer. Oriental hybrid pears include varieties that range from coarse and gritty to smooth, buttery textures. Some varieties have dessert, or fresh eating, quality that will rival the best European varieties. The more coarse textured varieties, such as Kieffer, are used primarily for home processing, including canning, preserves, pickled pears and baking.

The Asian pear, often termed "apple-pear," is gaining increased attention because of its unique fruit.

Pears are not self-fruiting, so you will need to plant more than one tree for pollination. Pollination is accomplished primarily by insect (mostly bees), so plant trees of different varieties within 40 to 50 feet of each other.

Common pear varieties

- **Anjou** is a large, broad, lopsided, greenish-yellow pear, which has brown speckles or russeting. There is also a red variety. A dual-purpose pear, it has juicy, sweet, somewhat spicy flesh. Anjou is the main winter pear of North America.
- **Bartlett** is the name used in North American and Australia, but in England it is known as English Williams. It is an irregularly shaped pear that is generally swollen on one side of the stem. Its speckled skin is golden yellow, with russet patches. Sometimes the skin is tinged with red. The tender, juicy flesh is creamy white; and the flavor is sweet, but slightly musky. Bartlett pears do not keep well. The season for these pears begins in late summer. Bartlett pears are excellent for fresh eating as well as canning and cooking.
- **Bosc** is a dark yellow, winter pear with a reddish cast to the skin. This variety is unmistakable with its long, tapering neck. The yellow-

ish-white, juicy flesh has an aromatic flavor. Very good for fresh eating as well as canning and cooking.
- **Comice** is one of the finest of pears, with creamy white, melting very juicy flesh and an aromatic flavor. The thick, yellowish green skin is covered with speckles and patches of russeting. These pears are larger than most, with a broad, blunt shape, and are available from late autumn to midwinter. Comice is an excellent pear for fresh eating.
- **Kiefer** pears are most likely a result of crossing an oriental pear with a Bartlett pear. Kiefer pears have coarse textured flesh that makes them less desirable as a fresh eating pear. They are most often used in cooking and canning. One reason for the popularity of this pear is its resistance to fire-blight.
- **Moonglow** is an oriental pear with good dessert quality. The medium to large fruits ripen in late summer. Moonglow has good fire blight resistance.
- **Seckel** is a very small dessert pear that has maroon and olive green skin. The ultra-sweet flesh of seckel makes it perfect for fresh

eating, although it also makes wonderful pickled pears.

- **Twentieth Century** is an Asian pear with good dessert quality, medium-size, yellow fruit. Best for fresh eating, Twentieth Century ripens in mid to late summer. It has Moderate fire blight resistance.
- **Warren** is an excellent dessert quality pear both in the flesh and peel qualities. Warren has a smooth texture and sweet flavor. It is a small to medium size pear with red-blushed skin. Warren ripens in late summer. It is highly resistant to fire blight.

Planting guidelines

When selecting the site for your pear trees, keep in mind the sun, moisture, and soil requirements of the trees. These requirements are basically the same as with apple trees.

Good moisture drainage is an important soil requirement although pears are more tolerant of poorly drained soil than most other fruit trees. Sandy soils are best, but pear trees can be grown in clay or heavy loam soils. Iron deficiency (chlorosis) and cotton root rot can be serious problems on highly alkaline soils. Iron chlorosis can be treated with soil applications of iron chelate.

Planting in full sun is a key factor for maximum fruit production. Choose an area that receives full or nearly full sun. Morning sunshine is particularly important for the reduction of disease incidence by the early drying of dew.

Pears bloom early, so the blossoms are often subject to spring freeze damage. This early freeze damage occurs most often on pears planted in low areas such as valleys, along streams, etc., or trees planted on the north side of buildings or slopes.

Plant pear trees in the winter or early spring while they are dormant.

Dig a planting hole large enough to spread the root system in a natural position. Do not add fertilizer to the planting hole.

Trim off any broken or mutilated root parts before planting. Set the plants at the same depth at which they were growing in the pots or at the nursery. Work soil in and around the roots, firming to eliminate air pockets as the hole is filled. Do not leave any depression around the tree. Water the tree thoroughly and check for air pockets. If the tree settles, gently lift it to the proper planting depth, adding more soil as needed.

Maintenance

Pruning a young tree controls its shape and strengthens it by developing a strong, well-balanced framework of branches. This framework is vital in fruit trees simply because of the weight of the crop. Branches that are not properly pruned will often split off the main trunk entirely if the fruit load gets too heavy.

Remove unwanted branches when they first develop to avoid the necessity of larger cuts in later years. Remove any crossed and rubbing branches to increase tree health.

Vigorous shoots are more vulnerable to fire blight, so if blight is a problem, be sure to use very little fertilizer. When the first new growth begins to appear, apply 1/2 cup of balanced fertilizer such as 13-13-13 in a 2 foot circle around the tree. Keep fertilizer at least 6 inches away from the tree trunk to avoid fertilizer burn. Each spring after the new growth appears, apply 1/2 cup of fertilizer such as 13-13-13 for every year of age of the tree through the fourth year. Thereafter, continue to apply about 2 cups of fertilizer per tree each spring after new growth develops.

Again, if fire blight becomes a problem, discontinue or drastically reduce fertilizer applications. If new growth amounts to less than 6 inches per year, increase the amount of fertilizer. Use ammonium sulfate instead of balanced fertilizer on highly alkaline soil (pH above 7.5) to avoid phosphorus-induced iron deficiency.

In dry climates, supplemental water may be required for optimal tree growth and fruit yields. Young trees should be deep watered at least once a week, as should older, more mature trees, however, older trees are more drought tolerant than young trees are. Fruit production and tree growth are always better if adequate water is available, despite tree age. Deep watering is achieved by applying enough water to soak into the soil several inches.

Pear trees will overbear in certain, favorable years, resulting in an abundance of smaller fruit as well as broken tree limbs. Removing excess fruit will encourage the development of larger size and more uniform shape of the pears remaining on the tree. The earlier thinning is completed, the more effective it is in achieving desired results.

Pick off developing fruit by hand. Leave only one pear per cluster and space the clusters approximately 6 inches apart. To remove fruit without damaging other pears on the spur, hold the stem between the thumb and forefinger and push the fruit from the stem with the other fingers. This method removes the pears but leaves the stem attached to the spur, preventing damage to the fruiting spurs.

Insect pests and diseases

Codling moth is the most serious insect pest of pears. Thinning fruit, trapping the insects, and appropriately timing insecticide applications can control this insect in home orchards. The larva of the codling moth is the worm responsible for "wormy pears." Wherever these fruit trees are grown, codling moth is a problem to some degree, and usually will need some form of treatment to protect the fruit.

Codling moth has two and maybe sometimes even three generations per year. Adult moths first emerge during mid-spring, usually around the period of full bloom. This flight continues for about six or seven weeks, peaking a couple of weeks after the first moths become active.

These moths lay eggs on the leaves of the trees. The larvae first feed on the leaves, later moving on to the developing fruit. After feeding within the fruit for a length of time, the full-grown larvae migrate out and crawl down the trunk of the tree to pupate under bark flaps or in other protected spots.

The second, and most important,

generation occurs in early summer. Adults emerge in early July, laying eggs directly on the fruit surfaces. The young caterpillars tunnel into the fruit shortly after the eggs hatch. Most of the fruit damage occurs at this time. Again, when full grown, these caterpillars leave the fruit to seek other protected areas to complete their development.

Insecticides are a common technique used to control codling moth and should be applied shortly after petal fall. Most homesteaders continue these treatments at 10- to 14-day intervals throughout much of the summer. Permethrin and carbaryl are the most common homeowner treatments for codling moth. Remember to never apply insecticides during periods of bloom. Treatments at this time will kill beneficial pollinators, such as honeybees.

Insecticide use can be limited and treatments better timed by using pheromone traps. These specialized traps are baited with the sex attractant of the female codling moth. The traps capture male moths and give an estimation of when mating and egg-laying take place. The first insecticide application often is optimal about three weeks after the first consistent captures of male moths. Pheromone traps that only collect males cannot control codling moth, although they can aid in pinpointing the proper timing for insecticide use

There are several non-chemical approaches that also help control codling moth.

Thinning fruits can limit codling moth infestations. Thinning also helps provide better control of codling moth when sprays are used in conjunction with thinning. Pick up and dispose of any fruit that show evidence of codling moth entry wounds.

A burlap or corrugated cardboard band around the trunk can cause many larvae to settle and pupate at these sites, where they can be regularly collected and destroyed.

Some control of adult codling moths is possible using homemade traps containing water and molasses mixtures, typically in about a 10:1 ratio. These water-molasses traps capture both sexes.

Scale insects sometimes cause damage to pear trees. Perhaps most damaging is the San Jose scale, a tiny, circular scale found on fruit as well as branches. Heavy infestations can cause death of infested branches and spotting of infested fruit. It can also contribute to a general decline of the tree.

Scale infestations are typically slow to

develop. Regular treatments are rarely needed in backyard orchards. Treatments with horticultural oils during dormant periods in early spring can provide adequate control of these insects.

Aphids will, on rare occasion, attack pear trees. On fruit trees, aphids over-winter as eggs clustered near the dormant buds. After the eggs hatch in spring, the aphids feed on the emerging leaves, causing he leaves to curl slight. Horticultural oils applied in late winter or early spring can prevent these problems. Diazinon, malathion and insecticidal soaps also can help suppress aphid outbreaks after leaves emerge. However, aphid infestations typically are short-lived on pears. Aphids usually leave the tree by late spring, flying to alternate summer host plants. If aphids are not seriously damaging the majority of terminals or injuring fruit, there is little benefit from insecticides for aphid problems after leaves emerge.

Beneficial insects such as lady bugs can also help keep aphid populations under control.

The pear slug is a rather unusual insect that feeds on the upper leaf surface of pears as well as some other fruit trees. The larvae look like slugs but are actually the immature stage of a type of non-stinging wasp called a sawfly. Feeding damage by pear slug larvae is highly characteristic in that injury is confined to areas between the main veins on the upper leaf surface. This produces a lacy, skeletonized injury on infested leaves.

Pear slug larvae are easy to control with any orchard insecticide, including insecticidal soaps. Larvae also may be washed off with a vigorous jet of water. A light dusting of the leaves with wood ashes is another highly effective, natural control.

Pear psylla are a serious pest of pears grown in the western United States. These aphid-like insects suck the sap from the leaves and produce large, sticky drops of honeydew that can coat large portions of the tree as well as the fruit. Heavy infestations greatly weaken pear trees.

Pear psylla spend the winter near previously infested pear trees. They return to the trees in early spring to feed and lay eggs. They can produce several generations each year.

Pear psylla are highly resistant to most available insecticides. Oil sprays, sometimes combined with an insecticide, can be effective when applied during the dormant season shortly after the pear psylla adults return to the trees. Sulfur-based sprays also can be effective but should never be applied with oils or plant injury may result. Pear psylla problems are greatly lessened in unsprayed orchards, where they are heavily attacked by beneficial insects.

Pears in the homestead orchard are commonly affected by several diseases every year. These diseases are scab, Fabraea leaf spot, sooty blotch, and occasionally fire blight.

Fire blight, cause by bacteria, is the most devastating disease of pears. Like most pear diseases, fire blight does not occur every year, however, when infection occurs, the disease develops quite rapidly and can destroy individual trees or even entire orchards in a single season.

The bacteria survive the winter in old cankers on pears and other plants and in healthy pear buds. This disease can occur in four phases: canker blight, blossom blight, shoot blight, and trauma blight. As the weather warms and becomes favorable for growth in the spring, the bacteria begin to multiply rapidly. A creamy, bacterial ooze can be seen on plant tissues. This ooze is attractive to insects. They pick it up and carry it to open flower buds where infection occurs. The bacteria are also carried by wind and rain to open pear blossoms. Infected tissues are characterized by their blackened, and burned appearance, hence the name "fire blight."

Prevention through selection of resistant varieties is the most effective means of control.

The second most effective method for control of this disease in home plantings is good sanitation. Any cankered or infected branches or twigs should be cut back to healthy wood during the dormant season. All pruning cuts should be made at least 8–12" below any visible symptoms. All tools should be disinfested with 10% bleach solution and then dried after each cut to prevent spreading this disease to healthy tree tissues. Prunings should also be removed from the vicinity of the tree, and preferably burned. In addition to these practices, it is important to be diligent in watching for new infections. Remove any blighted tissues as soon as they appear.

Fabraea leaf spot, also known as leaf blight and black spot, is caused by a fungus. This disease usually appears late in the growing season but can occasionally develop in late spring, around May or early June. Fabraea leaf spot attacks leaves, fruit, and twigs of pear. Symptoms first appear as brown to black spots on the leaves. Heavily infected leaves often yellow and drop prematurely. Severe defoliation can substantially reduce tree vigor and yield, especially if trees are defoliated several years in a row. Lesions on fruit appear similar to those on leaves but become slightly sunken as fruit expand. Severely infected fruit may also crack. Once established in a tree or planting, this disease is difficult to control since the disease can over-winter on infected leaves. Spores of the fungus are easily spread by rain and wind in the spring.

Effective control includes a good sanitation program. Since the disease often over-winters on infected leaves, removal of all fallen leaves during the dormant season can significantly reduce the opportunity for new infection. In addition, properly selected and timed fungicide sprays are important for disease control. Read fungicide label for proper use and application on pear trees.

Pear scab is a fungal disease that is quite similar to apple scab. The fungus causes circular, velvety, olive-black spots on leaves, fruit, and sometimes twigs. As the lesions age, they become gray and cracked. The fungus over-winters on dead, fallen leaves and produces spores in the spring that can infect the trees during periods of rain. Infection from these primary spores can take place anytime after pear growth begins until mid to late June if suitable weather conditions exist.

During the summer, a different spore is produced by the fungus that can begin more new infections when splashed onto leaves and fruits by rain.

This disease is effectively controlled by a good sanitation program in which diseased leaves and fruit are removed from the vicinity of the tree. This practice alone will significantly reduce sources of disease infection in the spring. Scab can also be controlled with properly selected and timed fungicide sprays. Read fungicide label for proper use and application on pears.

Sooty blotch is also caused by a fungus. It is recognized by the black, sooty smudges it produces on the surface of pear fruit.

Sooty blotch develops gradually during periods of high humidity, and is particularly severe when rainy weather occurs early in the season and continues into the summer. The fungus is favored by frequent showers, prolonged cloudy weather, and poor air circulation.

Since the fruit infections are superficial, they can usually be removed with vigorous washing and rubbing. In addition, practices that promote air circulation, such as proper, open pruning as well as mowing the grass around the tree, are usually enough to keep this disease in check.

Fungicide sprays can be applied if the tree has a history of severe disease and blemish-free fruit are an important feature such as fruit intended for sale.

Cotton root rot is another fugal disease that affects pears as well as many other plants. It is one of the most destructive plant diseases. The fungus is prevalent in certain clay loam soils primarily in the southwestern United States.

Disease symptoms are most likely to occur from June through September when soil temperatures reach 82 degrees F. The first symptoms of the disease are slight yellowing or bronzing of leaves followed by wilting. Usually, the fungus has done extensive damage to the trees roots by the time leaves have wilted. Under moist conditions, spore mats sometimes appear on the soil surface. These mats, up to a foot and a half in diameter, are first snow white and cottony, then later become tan and powdery.

The fungus generally invades new areas by a continual slow growth through the soil from tree to tree. The fungus can survive in the soil for many years, and often it is found as deep in the soil as tree roots penetrate. Affected areas often appear as circular areas of dead plants in fields of infected crops. These areas gradually enlarge in subsequent years as the fungus grows through the soil from plant to plant. Infested areas may increase 5 to 30 feet per year.

Cotton root rot is one of the most difficult plant diseases to control. At present, there is no effective treatment to eliminate this soil borne disease.

Anthracnose is a fungal disease that sometimes occurs in cooler climates. Anthracnose forms cankers that are often called fiddle-string cankers because of the long, string-like fibers of the inner bark that are left exposed in the center of the canker. These cankers usually occur on smaller twigs and branches and expand for only a single season. The fungus produces spores in these cankers, which are washed and splashed onto other branches and trees during rains or irrigation. Infection occurs most readily any time the weather is cool and damp, especially in the fall.

Harvesting

Most pears do not fully ripen on the tree. These pears are ready to harvest when they change from hard to firm (firmness that is similar to that of a softball). Harvest maturity is also usually indicated by a slight change from green to yellow. A few of the dessert type pears will fully ripen on the tree and remain good textured, sweet, and juicy.

Mature fruit will begin to drop even though still hard, if harvest is delayed. Most pears reach maturity in late summer with a few varieties ripening before or after that time. Pears should be picked and ripened off the tree. Pears remaining on the tree too long ripen poorly and lose quality in both texture and flavor.

Ripen harvested pears at room temperature in a well ventilated area. They will usually ripen in 1 to 2 weeks. Refrigerate the fruit after ripening until consumed or processed. For longer storage life, refrigerate unripe pears as near 32 degrees F as possible and then ripen as desired. Some pears, Bartlett for example, achieve the best quality when held in cold storage for a few weeks and then ripened.

Peaches and nectarines

Peaches and nectarines are basically the same, differing only in genes for surface fuzz. Therefore, a peach lacking the gene that contributes to peach fuzz is called a nectarine.

As with other types of fruit trees, peach rootstocks can influence tree size, productivity, cold hardiness, and tree longevity.

Dwarf rootstocks should be avoided

for peach and nectarine trees because some varieties die a mere 6 to 8 years due to incompatibility with the scion variety. Dwarf peach trees cannot usually handle the weight of a full fruit load and often it will only take a handful of fruit to split a tree, either killing it, or seriously damaging it.

There are more than 100 varieties of peaches and nectarines available from commercial nurseries. When choosing varieties for your homestead orchard, you should base the selection on climatic conditions, how you intend to use the fruit (fresh, frozen, canned, etc.), disease resistance, and season of ripening. Varieties requiring less than 800 hours of chilling often bloom and yield earlier, however, they are also more susceptible to early spring frosts.

Many varieties developed by Southern breeding programs have the shortest chilling requirements and are best adapted to southern growing conditions. There are also now a few varieties of peaches that can be successfully grown in all but the coldest of climates.

Most peaches and nectarines are typically self-fruitful and do not require cross pollination with other varieties. Therefore, trees of a single variety may be planted in large blocks. There are a few varieties that will not set a full fruit crop without another pollinizer planted close by. Most of these peaches are hybrids or crosses of other varieties. Check labels before purchasing your trees so that you can purchase pollinizers if necessary.

Peaches are available in freestone, semi-freestone (also called semi-clingstone) and clingstone. In a clingstone peach, the fruit adheres or clings to the stone. In a semi-freestone peach, the fruit can be pulled away from the stone when the peach is completely ripened. In a freestone peach, the fruit of the peach pulls freely away from the stone. The freestone varieties are the most popular for canning and freezing.

Common peach varieties
Yellow Flesh Peaches

- Laural is a medium sized, freestone peach that is well colored.
- Redhaven is a medium sized, semi-cling stone peach. Redhaven has delicious, juicy yellow flesh. It is also a very cold hardy tree.
- Contender is a large, attractive, freestone peach.
- Loring is a very large, firm freestone peach.
- Harcrest is a medium to large,

very attractive freestone peach.
- Encore is a large, firm, attractive freestone peach that is late ripening.
- Hale-Haven is a medium sized, oval peach that is noted for its sweet, juicy flesh. Hale-Haven is a cold hardy, high yielding, freestone variety.
- Elberta is homestead favorite with its juicy yellow flesh that is excellent frozen, canned or fresh. Elberta is a freestone variety.
- Golden Jubilee is a popular yellow freestone peach that is popular for its cold hardiness.

White Flesh Peaches
- Belle of Georgia is a large fruit with brilliant red coloring, very firm and highly flavored white flesh. Belle of Georgia is a freestone variety.
- White Lady is a medium sized, freestone peach with juicy, sweet, white flesh. It has good keeping qualities.
- Sugar Lady is a freestone peach with sweet white flesh. Sugar Lady is very vigorous tree producing high yields most years.
- Snowbrite is a small freestone

peach with outstanding color, and juicy, sweet, white-flesh. It is an early ripening peach.

Common Nectarine varieties
- Sunglo is a large, attractive, nectarine with good flavor. Sweet and juicy.
- Red Gold is a medium to large fruit with sweet aroma and good flavor
- Juneglo fruit are well colored and have good flavor.. This variety sets fruit in cool, wet weather and is known for high yields. Flesh is semi-freestone when ripe.
- Harko is a solid red, clingstone nectarine with good quality and flavor. It is a consistent producer with a tendency to overbear, so it must be thinned hard and early in order to get good fruit size.

Planting guidelines

When selecting the site for your orchard, keep in mind that cold air is heavier than warm air and cold air collects in low laying areas.

Peach and nectarine blossoms are fairly delicate and can be killed by air temperatures of 28 degrees F. Therefore, peach and nectarine trees should be planted on the tops or sides of hills at higher elevations than the surrounding area. Sometimes, just 10 feet in elevation can mean the difference between having or not having a crop.

Peaches and nectarines will thrive on a wide variety of soils. Because soil fertility can be easily adjusted, it is not a major consideration, however, soils with very high fertility should be avoided because trees grow too vigorously, producing lush foliage and low yields at a young age, and the fruit they produce is of poor quality.

Peach and nectarine trees perform best in full sun, and where the wind is not too vigorous. Peaches and nectarines do not like standing water so well drained but moisture rich soil is a must.

Plant your trees in the spring in holes dug slightly deeper and wider than the size of the pot or root ball in which they were purchased. Be sure to keep the bud union 1 inch above the soil line. Planting a peach or nectarine tree too deep in the soil can cause poor growth or even death of the tree. Back-fill and firm the soil around each tree. Water the new trees well after planting.

Immediately after planting, prune the tree back to a height of 26 to 30 inches. Cut off all side branches to leave a whip (a shoot without any side branches or with all side branches removed). Ideally this whip should be about 26 to 30 inches tall. This may sound drastic, however, the healthiest, best shaped, open-center trees come from those that were pruned initially to a whip.

Apply 1/2 pound of 10–10–10 fertilizer or its equivalent 7 to 10 days after planting to each tree, and the same amount again to each tree 6 weeks after planting. Broadcast the fertilizer evenly, 8 to 12 inches away from the trunk.

Maintenance

During the first year, remove any diseased, broken, and low-hanging limbs. Also remove vigorous upright shoots that may have developed on the inside of the main tree scaffold, that if left could shade the center of the tree.

During the second and third years, remove low-hanging, broken, and diseased limbs. To maintain the open vase, remove any vigorous upright shoots developing on the inside of the tree, leaving the smaller shoots for fruit production. Finally, prune the vigorous upright limbs on the scaffolds by cutting them back to an outward growing shoot.

Remember, the more severe the angle of a crotch, the more likely it is to split under a heavy fruit load. Wide angle crotches are the healthiest and strongest. Consider it to be the fingers of your hand. Thumb crotches are good, finger crotches are bad.

The same principles used to develop the trees are subsequently used to annually maintain the size and shape of the mature tree. Remove any low-hanging, broken, and dead limbs first. Next, remove the vigorous upright shoots along the scaffolds. Lower the tree to the desired height by pruning the scaffolds to an outward growing shoot at the desired height.

In years when no frost or freeze damage affects the spring blossoms, more fruits will set than the tree can support and fruit must be thinned. Approximately three to four weeks after bloom or when the largest fruit are as large as a quarter, fruits should be removed by hand so that the remaining peaches are spaced about every 8 inches. Fruit thinning will allow the remaining fruits to develop optimum size, shape, and color, and prevent depletion of tree health. Pick off misshapen fruits, fruits with visible insect damage, or unusually small fruits first.

In the second and third years after planting, each tree should receive 3/4 pound of 10–10–10 in March and again in May. Mature peach and nectarine trees between 4 and 10 years of age should receive 1 to 2 pounds of 10–10–10 fertilizer each in March and May. If the tree is vigorous and there is little or no fruit expected that year, only the March application is necessary. Broadcast the fertilizer around the outer edge of the tree keeping the trunk area free of fertilizer.

Peach and nectarine trees need to achieve 18 inches of new growth each year. Remove any sod that grows under the trees, mulch and irrigate as needed. Irrigation will increase yield particularly if it is applied three weeks before harvest, especially in more arid climates.

Thinning is an important step in maintaining healthy peach and nectarine trees. Thinning of the developing fruit is done to help prevent splitting and breaking of branches under a heavy fruit load, as well as to help produce larger, more flavorful fruits.

When thinning peaches or nectarines, thin the fruits at two stages during the growing season. First in late May when the fruits are beginning to form, the again in mid-July when the fruits are about half their full size. During the each thinning remove any damaged or diseased fruit. Then, thin the fruit so that there is enough room for the peaches or nectarines to develop without touching each other leaving about 4 inches between each fruit.

If the weight on some of the branches looks like it may cause them to break, it will be necessary to support the branch with a T-brace from below. A couple of boards nailed into a T shape is all that is needed to produce the brace. Where this is done, padding will need to be added to the arm of the brace where the brace meets with the branch. This will prevent the branch rubbing on the stake and possibly creating an entry wound for insects or disease.

Insect pests and diseases

Numerous insects attack peach and nectarine trees. These insects can cause extensive damage to the blossoms, fruit, leaves, twigs, limbs and trunk. Some of the most common insect pests of these fruits are plum curculio, Oriental fruit moth, Peachtree borer, lesser Peachtree borer, shot hole borer, catfacing insects, scale, Japanese beetle and the green June beetle.

The adult plum curculio is a mottled brown beetle with a rough and warty body surface. It is about ¼ inch long and has a long, curved snout. Its immature stage is a grub (larva). A fully mature grub is legless, smooth-bodied and up to ½ inch in length. It is yellowish to grayish in color and slightly curved with a brown head.

Both adults and grubs cause damage to peaches, as well as plums and other stone fruits. The primary injury is caused by the adult female when she makes a crescent-shaped cut in the skin of the fruit to lay her eggs. This results in D-shaped scars on the fruit surface. Grubs that hatch from the eggs feed in the fruit, utterly ruining it. Later in the season both males and females damage the fruit by making round feeding punctures.

Plum curculio adults over-winter under leaves, brush and in other protected places near the orchard. Wild plum thickets within ¼ mile of an orchard can provide a source of infestation. They usually become active at about the time the trees bloom. The adults will feed on developing fruit and leaves and then lay their eggs in the young fruit.

Carefully inspect young fruit for egg-laying and feeding scars. Fruit infested shortly after bloom by the first generation usually drop to the ground. The larvae then hatch and feed in the fruit. After developing, they leave the fruit, burrow into the ground and pupate During mid-summer the first generation adults emerge, move into the trees, and begin laying eggs. Fruit infected by the second generation remain on the tree until harvest. The second generation adults emerge in the fall, move to the hibernation areas and over-winter.

When disturbed, the adult plum curculio tends to fold its legs against its body and fall to the ground where it remains motionless for several minutes. Place a sheet or other light-colored cloth on the ground under the tree and shake some branches. If present, the plum curculio will drop to the ground and be readily visible.

Good sanitation is vital in controlling this pest. Pesticides should be applied after the petals fall to aid in controlling the first generation of the beetle. Three sprays, the first in mid-June, and then two more sprays spaced about two weeks apart will control the second generation. Carbaryl is an effective pesticide for use against this pest. Carbaryl can also be used up to 3 days prior to harvest if necessary. The pre-mixed pesticide spray products sold for home orchards will also be effective controls. As with all pesticides, read and follow all label directions and precautions.

The Oriental fruit moth is grayish-brown and has a wingspan of about ½ inch. Like most moths, it is active at night.

When first hatched, its larva is tiny, and white with a black head. The mature larva is larger, has six distinct legs and is pinkish with a brown head. The Oriental fruit moth is a pest of peaches and other stone fruits. In some areas, there are six or more generations of Oriental fruit moth per year. This pest over-winters as mature larvae inside cocoons, which are located in protected areas on the tree or in debris near the base of the tree. In early spring, the larvae pupate and adults begin to emerge around the time of bloom. The adults then lay eggs from which more larvae hatch. These first-generation caterpillars bore into new growth at the tips of peach tree branches. This activity causes the branch tips to wilt and die back. Later in the season, after the branch tips harden, caterpillars enter the fruit instead and feed on them.

While in the fruit and twigs, caterpillars are protected from insecticides. Good early season control of adult moths using proper sanitation methods and insecticides will often provide control for the entire season.

The presence of Oriental fruit moths can be detected with the use of traps containing pheromones that will attract the adults. Permethrin sprays should be applied if an average of more than 10 moths per trap occurs. Do not apply permethrin within 14 days of harvest, and do not apply more than 8 applications of permethrin per season for all insect pests.

The Peachtree borer, the lesser Peachtree borer, and the shothole borer are serious pests of peaches and nectarines. Of these, the Peachtree borer and lesser Peachtree borer are the more destructive pests. They are found on most cultivated and wild stone fruits, including some ornamental shrubs such as flowering peach and cherry. It is the larvae of these insects that causes so much damage to these fruit trees.

The peachtree borer adults are clearwing moths, and are often mistaken for wasps because of their appearance and behavior. The adult female peachtree borer is a metallic blue-black color except for a reddish band on the abdomen. The male is black with yellow stripes along the back at the base of each wing and narrow yellow stripes on the abdomen. The larva is about an inch long when fully grown. It is a creamy white color with a brown head.

The larva of the peachtree borer attacks the base of the trees and may be found feeding from the main roots to about 10 inches up onto the trunk. Masses of gum mixed with frass (a sawdust-like insect waste) are the primary symptoms of attack by peachtree borers. Young trees can be killed by a very small number of larvae. Older trees can tolerate more larvae but often succumb to this pest.

The peachtree borer over-winters as larvae. It has one generation per year. Some adults begin emerging in late May although peak emergence is in mid- to late August. Wounds and rough bark are favorite sites for egg laying. About two weeks after the eggs are laid at the base of the tree, the small larvae hatch, burrow into the bark and begin feeding. They stop feeding when cold weather comes and resume feeding the following spring.

Since the peachtree borer causes its most severe damage to young trees, extra care must be taken during planting to avoid damaging the bark. A pre-plant dip in an insecticide solution is strongly recommended. Annual trunk sprays during July or August will generally keep the peachtree borer under control. Be sure to apply sufficient spray from the scaffold limbs to ground level so the bark is saturated and a small puddle forms at the base of each tree. Homesteaders can use permethrin for insecticidal control. Do not apply permethrin within 14 days of harvest.

The adults of the lesser peachtree

borer are also clearwing moths. Both the male and female adult lesser peachtree borers resemble the male peachtree borer except that they are somewhat smaller. The larva of the lesser peachtree borer is very similar to the larva of the peachtree borer but smaller.

The lesser peachtree borer attacks the trunk and main limbs. Again the symptoms are oozing gum that contains frass. Heavy infestations can kill individual limbs or an even an entire tree.

Like the peachtree borer, the lesser peachtree borer over-winters as larvae. It, however, has two generations per season and occasionally, a third. Emergence of adults peaks in late April to mid-May and late July to mid-August. *Cytospora* canker, fungal disease, new wounds, and previously infested areas are favorite sites for egg laying.

The best control for the lesser peachtree borer is to keep the trees in a vigorous, healthy condition and to prevent any mechanical injury. Proper pruning of any damaged, diseased, or insect infected limbs is an important step in controlling this insect. Destroy pruned wood before adults emerge in April by burning. Avoid spreading bacterial canker while pruning by dipping the pruning tool after each cut into a solution of one part household bleach to nine parts water.

As with the peachtree borer, annual trunk sprays in late July or August will help control the lesser peachtree borer. However, since there are two or more generations per year it is difficult to get good control with insecticides since the first generation emerges while there is fruit on the tree. Homeowners can use permethrin, Do not apply permethrin within 14 days of harvest.

Shothole borers are small, cylindrical beetles that attack many kinds of fruit trees as well as ornamental trees and shrubs. Plants under stress are highly susceptible to shothole borer attack.

Shothole borers attack the trunk and limbs. The entry holes look like the tree has been hit with bird shot. The adult beetle bores into the bark and then carves out chambers below the bark in which to lay its eggs. The larvae then hatch and feed on the bark of the trees.

Shothole borers over-winters as larvae. It has several generations per year. The adults emerge from the infested trees in April and May and move to new trees, especially those under stress from drought, or disease.

The best control for shothole borer is to keep the trees in a vigorous, healthy growing condition and to prevent any mechanical injury that might make a point of entry for the pest. Prune out split or broken limbs and limbs with signs of borer damage whenever possible. Destroy pruned wood before adults emerge in April by burning. Avoid spreading bacterial canker while pruning by dipping the pruning tool after each cut into a solution of one part household bleach to nine parts water.

Pesticide sprays that are effective for other insects usually provide adequate control of shot hole borer adults As stated before, do not apply permethrin within 14 days of harvest. With more than one generation of borers per year it is difficult to get good control with insecticides since the first generation emerges while there is fruit on the tree. There is no effective control for insects that are already in the tree.

Catfacing insects include the tarnished plant bug, and various stink bugs.

The tarnished plant bug is oval and has a white triangle high on its back. It is brown and about ¼-inch long.

Stinkbugs are shaped like a shield. They vary in color from green to brown and in size from ½ to ¾ inch in length.

The tarnished plant bug and the stink bugs have needlelike mouthparts that they use for piercing and sucking. They distort fruit by their feeding. The damage that they cause appears as deep dimples in the fruit. The damage is cosmetic and the fruit is still edible.

The catfacing insects over-winter as adults in protected areas in or near the orchard. Annual weeds that begin to bloom in late winter or early spring are a major attractant for these insects.

Good sanitation is important for controlling cat facing insects. Removing weeds and debris in the area around your orchard will greatly reduce the incidence and damage of these pests. For chemical insecticide control, carbaryl or permethrin are equally effective. Do not apply permethrin within 14 days of harvest, or carbaryl within 3 days of harvest.

Various scale insects also attack peaches and nectarines. One of the most commonly seen of these scale insects is the white peach scale.

Scales are unusual in appearance in that they are small and immobile, with no visible legs. Scales vary in appearance depending on age, sex and species. The adult females typically produce a waxy covering that protects them from many insecticides. They feed on sap by piercing the leaf or stem with their mouthparts and sucking.

The adult female white peach scale is

tiny, no more than 1/8th inch in diameter. It is circular in shape and yellowish to grayish white with a yellow or reddish spot.

White peach scale will infest the bark, fruit and leaves of peach and nectarine trees. An infestation by this scale can result in stunting, leaf drop, and the death of branches or even the death of entire trees.

The white peach scale survives the winter as an adult female. The adult male is mobile and only lives about one day. After mating, the female starts laying eggs in mid-spring. The eggs hatch into nymphs. These nymphs crawl around for a few days before settling in and beginning to feed. There can be up to three generations per year in warmer climates.

The adult female scales are difficult to control with insecticides because of their hard, waxy covering. Dormant oil can be applied before buds break in late winter or early spring. Dormant oil works by smothering the over-wintering adult females. Chemical control of the crawlers can be achieved with carbaryl or malathion. Sprays for controlling white peach scale should be applied at bud bread, and again 2 months later, then again in another 2 months in areas with heavy infestations. Do not apply malathion within 7 days of harvest, or carbaryl within 3 days of harvest. As with all pesticides, read and follow all label directions and precautions.

Brown rot is one of the most common and serious diseases affecting peaches and nectarines. Brown rot is caused by a fungus that infects fruits, flower blossoms and shoots. The disease begins at bloom. Infected flowers wilt and turn brown very quickly. Shoot infections are usually a transference of the fungus from blossom infection. It results in small gummy cankers, less than 3 inches in diameter. Spores from infected flowers and cankers go on to infect fruits during wet periods. Diseased fruits remain attached to the tree and provide an additional source of spores for more infections instead of dropping off the tree in a normal fashion. Infections in apparently healthy green fruit remain inactive until the fruit begins to ripen.

Fruit rot starts with a small, round brown spot, which expands to eventually rot the entire fruit. Infected fruit turns into a mummy on the tree. The fungus survives the winter on fruit mummies (on the tree and on the ground) and twig cankers.

To help control Brown rot, collect and remove diseased fruit from the tree as it appears. In the fall remove all dried fruit mummies from the tree, since this is where the fungus sill survive the winter. Fungicides are usually required if fruit ripening occurs during a period of warm, wet weather. It is important to begin spraying just before the fruit begins to ripen. Look for the first tinge of change in the yellow background color. Prevention is much more effective than control once the disease appears on your trees. Fungicides containing thiophanate methyl, captan, or azoxystrobin that is labeled for use on peaches is most effective against this disease. These fungicides are only effective if complete and thorough coverage of the trees can be obtained. Always apply all pesticides according to directions on the label.

Peach scab, also known as "freckles," is caused by a fungus. Symptoms of Peach scab are small, velvety dark spots and cracks on the fruit. In cases of severe infection, spots may join together to form large dark lesions. Leaf infection is usually not observed. Twig infections occur on the current year's growth and are light brown after 30 to 70 days, before later enlarging and becoming dark reddish brown the next season. Spots on the fruit only occur on the outer skin, and eating quality is not affected. Peeling the fruit removes any sign of this disease.

Most peach and nectarine varieties are susceptible to scab, although some are more severely affected than others are. Minimize infection by selecting planting sites that are not low-lying. Trees should be properly pruned to allow good air circulation. This helps to promote rapid drying of the leaves, fruit, and twigs and minimizing the opportunity for fungal infections.

Periods of rain with temperatures of 65 to 75 °F are optimal conditions for scab infection. Fungicides can provide adequate control of this disease if applications are properly timed. Apply either wet-able sulfur or an appropriately labeled fungicide. Make five applications after petal fall at 7 to 14 day intervals. Always use and store all chemicals according to label directions.

Bacterial spot is a disease that affects peach and nectarine fruit and leaves. Infected leaves develop small reddish-purple spots that often have a white center. In advanced cases, the inner portion of the spot often falls out, giving the leaf a "shot-hole" appearance. Infected leaves turn yellow and drop from the tree. Lesions on fruits appear as small dark spots, which become larger and crater-like as the fruit grows. These lesions are generally shallow but can be up to ¼ inch deep. They do not develop the velvety spots of scab. Peeling the fruit will remove most traces of this disease.

Bacterial spot is difficult to control. Varieties are available that are moderately resistant, but not immune. Bacterial spot is usually more severe on poorly nourished trees or where nematodes are a problem, so proper fertilizing and orchard sanitation is important.

Peach leaf curl is caused by a fungus, and can infect peach and nectarine leaves, flowers, and fruit. Infected leaves pucker, thicken, curl and often turn red. These infected leaves eventually turn yellow and drop off from the tree. Severe leaf drop can weaken the plant and reduce fruit quality. Fruit symptoms are raised, wrinkled areas that are often overlooked by the homesteader.

Control is impossible after the symptoms become visible. Fungicides applied before bud break usually give good control. A single dormant application is entirely sufficient. This application may be mixed with spray oil for added scale and mite control.

If peach leaf curl infection has been severe enough in the past to warrant chemical control, choose an appropriately labeled fungicide, usually copper based, and apply and store according to label directions.

Gummosis is a fungal disease that can kill branches or even entire trees. Earliest symptoms appear on the young bark of vigorous trees as small blisters, usually occurring at lenticels. Infection occurs late in the season, and may be apparent in the fall or the following spring. Some infected areas exude a gummy resin. Trees that are two or three years old, often have sunken diseased areas, called cankers, that are apparent on the trunk and major branches. Large amounts of gummy resin, or gum balls, are associated with these lesions at multiple sites. After repeated infections, the bark becomes rough and scaly.

There is no practical chemical control available. Prevention is best and keeping trees healthy is the best prevention, since the most severely infected trees are water-stressed. Dead wood should be removed during winter pruning, and destroyed immediately. Where gummosis is present, use of captan for scab control is the preferred treatment.

Powdery mildew is a fungal disease that is primarily a problem on green fruit, but can also occur on leaves and young shoots. It appears as a powdery white coating on infected surfaces, and new shoots and leaves may be distorted. Young fruit develop white, circular spots that may enlarge. Infected areas on fruit turn brown and appear rusty. Symptoms usually occur on green fruit and disappear as the fruit matures.

As with most fungi, providing good air circulation to the trees by thinning and following proper pruning practices will help prevent and control this disease. Good orchard sanitation is also an important step in control. This disease occurs frequently when roses are nearby.

Crown gall is a disease caused by soil-inhabiting bacteria. The symptoms are rough, rounded galls or swellings that occur at or just below the soil surface on stems or roots. Young galls are light green or nearly white in color. As they age, the galls become darkened and woody and range in size from small swellings to several inches across. The galls disrupt the flow of water and nutrients traveling up from the roots and stems, thus weakening and stunting the top of the plant.

Occasionally, the disease becomes systemic and the galls are seen above the ground as well. Crown gall affects other ornamental plants as well, so use care in growing ornamentals in close proximity to your orchard.

There are no chemical controls available for crown gall in the homestead orchard. For new plantings select disease-free trees that have no evidence of galls. The bacteria enter through fresh wounds, so avoid injury to the roots and crown during planting and cultivating. Remove infected trees as soon as any galls are observed. Disinfect all cutting and pruning tools that have been used anywhere near crown gall. To disinfect tools, dip them for several minutes in a solution of one part household bleach to nine parts water.

Root and crown rots are very important diseases that affect stone fruits. Trees often die within weeks or months of the first symptoms, but in other cases the decline is gradual, occurring over several growing seasons. These diseases are caused by fungi, and are most severe in areas of poor drainage. Infected trees have stunted shoot growth and the leaves become sparse, small and yellowed. Fruit will be small and appear sunburned. Shoot and scaffold limbs die back as the disease progresses.

Crown rot symptoms appear as black decayed areas on the root crown and/or trunk base near the soil line. Cankers that exude a gummy resin are often present.

Root rot symptoms include few feeder roots being present with the remaining roots often badly decayed.

There are no chemical controls available for crown and root rot in the homestead orchard. The most important control strategy is careful water management. Do not overeater trees. Select well-drained sites for planting, and improve drainage of the existing location if necessary.

Oak root rot is also caused by a fungus. Initially trees infected with oak root rot appear slow in growth rate, have shorter terminals and take on an off-color green. As the root rot gets closer to the root crown, the whole tree or significant portions of tree can collapse. This collapse can occur at anytime during the year. There are no root sprouts present. Removing the bark beneath the soil surface reveals a white mantle of fungus between the bark and the wood. The wood remains firm and intact.

There is no treatment or prevention once the tree is in the ground. Do not plant peach or nectarine trees where oak trees have been removed.

Peach tree short life is a disease caused by the ring nematode, bacterial canker organism. Contributing to this disease are fluctuating winter temperatures, pruning the wrong time of year and poor horticultural practices. Trees suddenly collapse shortly after leafing-out or prior to leafing-out in the spring of the year. Removing a piece of bark from the lower trunk has a characteristic sour sap odor. The root system appears healthy and frequently puts up a flush of sprouts.

Prune peach and nectarine trees only in February and early March. Adjust the soil pH to 6.5 prior to planting and lime regularly to maintain this pH after planting if necessary. Select sites that are on heavier soils and are well drained. There is no effective method of nematode control after planting for the homestead orchard. To help prevent this disease, select peach trees that use the variety 'Guardian' for their rootstock. 'Guardian' is more tolerant of the ring nematode.

Do not replant old peach tree sites with new peach trees. Where ring nematode is present plant Stacey wheat as a winter crop and sorghum as a summer crop at least one year in advance of planting your orchard and two years is preferred. Fertilize to maintain at least 18 inches of new terminal growth per year. Remove all dead wood and dying branches as soon as possible and destroy immediately.

Harvesting peaches and nectarines

Pick peaches and nectarines after the green tinge completely disappears. Only then will the fruit taste ripe and sweet. Peaches should be slightly firm, but should come off the branch fairly easily with just a slight twist. Harvest peaches carefully to avoid bruising the tender flesh.

Use peaches and nectarines within a day of picking, if possible. Otherwise, store them in the refrigerator's crisper for up to three to five days. Can, dry or freeze peaches for long-term storage, or make them into preserves, jam, pickles or salsa.

Plums

Plums are in the family of stone-fruits, and are related to cherries, nectarines, and peaches. Many aspects of their growing and care are similar other stone-fruits.

The key to growing plums is to select a variety that is suitable to your zone. There are two major types of plums; European and Japanese.

European plums are generally hardy in zones 5–9 and Japanese plums are considered hardy in zones 6–10.

Certain American plums and hybrids are extremely cold hardy and drought tolerant and are good as far north as zone 4. Select disease resistant cultivars whenever possible. Most European and American plum varieties are self-fertile, but produce much higher yields if planted with other cultivars.

Japanese plums must be cross-pollinated, either by American or other Japanese cultivars.

A mature, full-size plum tree can produce over 50 pounds of plums every year, so you should seriously consider uses when deciding upon rootstock and the variety of plum you choose to grow.

Some plum trees are self-fertile, but many require a compatible plum tree nearby for pollination to occur. Plum trees have a short and very distinct pollination period so if you choose a tree which is not self-fertile, be sure to also choose a compatible pollinizer tree. This is especially important if you choose a 'gage' type of plum.

Common plum varieties
- Bullace is strictly a cooking plum. The trees are smaller than normal and very hardy, they are also ornamental. The fruit has a very sharp or tart flavor, excellent for jams and preserves.
- Damson plums, or Damask plums, are identified by their small, oval shape that is slightly pointed at one end. The fruit has smooth-textured yellow-green flesh,

and skin that ranges from dark blue to indigo. Damsons are usually harvested in late summer.

The skin of the damson is heavily acidic, causing the fruit to be rather unpalatable for eating out of hand, although they are somewhat sweeter than Bullace. Because of this acidic, tart flavor, damsons are commercially grown for use in cooking preserving, such as use in jams and conserves, etc.

- Gages are eating, or desert plums. They are some of the sweetest plums available and they have a very distinct fragrance. Gages are known for their rich, confectionery flavor that cause them to be considered one of the finest dessert plums. They are identified by their small, oval shape, and smooth-textured flesh. They range in color from green to yellow.

Planting guidelines

Select a planting site with full sun exposure and average to rich, well-drained soil that has a slightly acidic pH (6.0 to 6.8). Avoid low-lying areas that are prone to frost or areas with standing water and sites where cherry, peach or plum trees have been grown in the past.

Japanese varieties prefer well-drained soil that is rich in nutrients, while European varieties are more tolerant of heavy clay soils.

Plant plum trees in the spring or fall. Trees should be set to that the graft union is an inch or so higher than the soil line. Space standard-size trees from 20 to 25 feet apart and dwarf varieties from 15 to 20 feet apart. Dig hole slightly larger and deeper than the pot or root ball of the new tree. Set tree as straight as possible in the hole. Backfill with rich soil and tamp down firmly around root ball. Water tree well. Be sure to provide adequate moisture for the first few months that the tree is in the new location.

Maintenance

Plums are thinned for the same reasons other fruits are thinned, to prevent branches from splitting and breaking under a heavy fruit load, and to produce larger, more flavorful fruits. When thinning plums, thin the fruits at two stages during the growing season. First in late May when the fruits are beginning to form, the again in mid-July when the plums are about half their full size. During the each thinning remove any damaged or diseased fruit. Then, thin the fruit so that there is enough room for the plums to develop without touching each other leaving about 3 inches between each plum.

Plum trees are best trained to an open center for a wide-spreading tree. Plum trees should be pruned in much the same fashion as peach trees are, but they should have more usable branches and these should be pruned more lightly than the peach. Plum trees tend to develop open centers naturally without the precise procedures used in pruning other fruit trees such as apples and peach trees. With both young and mature plum trees, you'll want to remove all branches that interfere with basic framework of the tree, such as any branches that hang down below the lowest common point, branches that rub against another, etc. Also remove branches that have weak, narrow crotch angles. Remember to use your hand as a guide. Thumb crotches are good, finger crotches are bad. When branches are growing closely parallel, remove the weaker of the two, or the one less desirably located.

Plum trees should be pruned yearly. Remove any crossing branches and those that are growing into the center. Be sure to cut any new downward growing branches to maintain symmetry and strength. Do a fair amount of thinning out of any crowded areas of the tree.

During the growing season, rub off all suckers from the main branches within two feet from the trunk and all suckers sprouting from the rootstock. Most of the pruning should be done in late winter, but topping of vigorous shoots may be done in early summer.

Insect pests and diseases

Plum trees are relatively pest-free fruit trees, but check them regularly for signs of fruit pests and disease.

Some of the most damaging insect pests to attack plums is the peach tree borer, the lesser peach tree borer and the shot hole borer. Of these, the peachtree borer and lesser peachtree borer are the more destructive pests. They are found on most cultivated and wild stone fruits, including some ornamental shrubs such as flowering peach and cherry. It is the larvae of these insects that causes so much damage to these fruit trees.

The peachtree borer adults are clear-wing moths, and are often mistaken for wasps because of their appearance and behavior. The adult female peachtree borer is a metallic blue-black color except for a reddish band on the abdomen. The male is black with yellow stripes along the back at the base of each wing and narrow yellow stripes on the abdomen. The larva is about an inch long when fully grown. It is a creamy white color with a brown head.

The larva of the peachtree borer attacks the base of the trees and may be found feeding from the main roots to about 10 inches up onto the trunk. Masses of gum mixed with frass (a sawdust-like insect waste) are the primary symptoms of attack by peachtree borers. Young trees can be killed by a very small number of larvae. Older trees can tolerate more larvae but often succumb to this pest.

The peachtree borer over-winters as larvae. It has one generation per year. Some adults begin emerging in late May although peak emergence is in mid- to late August. Wounds and rough bark are favorite sites for egg laying. About two weeks after the eggs are laid at the base of the tree, the small larvae hatch, burrow into the bark and begin feeding. They stop feeding when cold weather comes and resume feeding the following spring.

Since the peachtree borer causes its most severe damage to young trees, extra care must be taken during planting to avoid damaging the bark. A pre-plant dip in an insecticide solution is strongly recommended. Annual trunk sprays during July or August will generally keep the peachtree borer under control. Be sure to apply sufficient spray from the scaffold limbs to ground level so the bark is saturated and a small puddle forms at the base of each tree. Homesteaders can use permethrin for insecticidal control. Do not apply permethrin within 14 days of harvest.

The adults of the lesser peachtree borer are also clearwing moths. Both the male and female adult lesser peachtree borers resemble the male peachtree borer except that they are somewhat smaller. The larva of the lesser peachtree borer is very similar to the larva of the peachtree borer but smaller.

The lesser peachtree borer attacks the trunk and main limbs. Again the symptoms are oozing gum that contains frass. Heavy infestations can kill individual limbs or an even an entire tree.

Like the peachtree borer, the lesser peachtree borer over-winters as larvae. It, however, has two generations per season, and occasionally, a third. Emergence of adults peaks in late April to mid-May and late July to mid-August. *Cytospora* canker, fungal disease, new wounds, and previously infested areas are favorite sites for egg laying.

The best control for the lesser peachtree borer is to keep the trees in a vigorous, healthy condition and to prevent any mechanical injury. Proper pruning of any damaged, diseased, or insect infected limbs is an important step in controlling this insect. Destroy pruned wood before adults emerge in April by burning. Avoid spreading bacterial canker while pruning by dipping the pruning tool after each cut into a solution of one part household bleach to nine parts water.

As with the peachtree borer, annual trunk sprays in late July or August will help control the lesser peachtree borer. However, since there are two or more generations per year it is difficult to get good control with insecticides since the first generation emerges while there is fruit on the tree. Permethrin is effective, however, do not apply permethrin within 14 days of harvest.

Shothole borers are small, cylindrical beetles that attack many kinds of fruit trees as well as ornamental trees and shrubs. Plants under stress are highly susceptible to shothole borer attack.

Shothole borers attack the trunk and limbs. The entry holes look like the tree has been hit with bird shot. The adult beetle bores into the bark and then carves out chambers below the bark in which to lay its eggs. The larvae then hatch and feed on the bark of the trees.

Shothole borers over-winters as larvae. It has several generations per year. The adults emerge from the infested trees in April and May and move to new trees, especially those under stress from drought, or disease.

The best control for shothole borer is to keep the trees in a vigorous, healthy growing condition and to prevent any mechanical injury that might make a point of entry for the pest. Prune out split or broken limbs and limbs with signs of borer damage whenever possible. Destroy pruned wood before adults emerge in April by burning. Avoid spreading bacterial canker while pruning by dipping the pruning tool after each cut into a solution of one part household bleach to nine parts water.

Pesticide sprays that are effective for other insects usually provide adequate control of shot hole borer adults As stated before, do not apply permethrin within 14 days of harvest. With more than one generation of borers per year it is difficult to get good control with insecticides since the first generation emerges while there is fruit on the tree. There is no effective control for insects that are already in the tree.

Green peach aphid and mealy plum aphid commonly curl the new growth of stone fruit trees, including plums. These insects sometimes seriously damage the tree. Although leaves produced later, after the insect has left the plant, are healthy, the first flush of leaves may be so curled and distorted that they die or provide little food energy to the tree, reducing fruit production. Dormant oil spray used before bud break and again about 2 weeks later is usually an effective control. Once the leaves show signs of aphid damage, there is no effective control.

The plum gouger is a weevil that is a particularly common insect pest in the hard European plum varieties. It spends the winter in the adult stage around the base of previously infested trees. In spring, about the time blossoming occurs, the plum gouger moves back into the trees to feed on the buds and flowers. Later they feed on small fruit, producing puncture wounds that typically ooze sap. Eggs are laid in some of the fruit and the young grubs develop on the pit of the fruit.

Effective controls have not been identified for these insects. They will often drop off from the foliage and fruit when the tree is disturbed, so some beetles can be collected by shaking branches over a sheet during blossoming and fruit set. Fruit tree sprays applied before and after bloom should also aid in controlling them.

Tent caterpillars are a cyclical pest of stone fruits. They over-winter in the egg stage. These eggs hatch into caterpillars in spring and early summer. Tent caterpillars have one generation each year.

Damage caused by tent caterpillars may be serious on individual trees. From April to June both Eastern and Western tent caterpillars build large silken tents over leaves on which they feed. Forest tent caterpillars build mats of webbing rather than tents. They forage in all directions from these mats or tents but return to the colony when not feeding. Tent caterpillars do not eat the leaf veins but will skeletonize leaf systems.

Populations of tent caterpillars tend to be concentrated in individual trees scattered throughout the orchard. Treatment is only occasionally required and can be limited to small areas of the orchard.

Bacillus thuringiensis (Bt) sprays, along with pruning out infestations are organically acceptable management methods.

On small trees, cut out and destroy infested twigs. Spray programs for other insects generally reduce tent caterpillar populations. If insecticide treatments are required, localized treatments on individual trees and infested branches are generally all that is necessary. Treat when small caterpillars are first observed. The addition of a wetting agent to increase penetration of the webbing by the insecticide enhances control.

Black knot is a fungal disease of plums and cherries. It is a widespread and serious disease throughout most of the United States. The disease becomes progressively worse during each growing season and unless effective control measures are taken, it can stunt or kill the tree. The black knot fungus can infect American, European, and Japanese varieties of cultivated plums and prunes. It also affects sweet as well as sour cherries, and peaches and nectarines on occasion as well.

The black knot fungus mainly affects twigs, branches, and fruit spurs. Occasionally, trunks may also become diseased. Usually, infections originate on the youngest growth. On infected plant parts, abnormal growth of bark and wood tissues produce small, light-brown swellings that eventually rupture as they enlarge. In late spring, the rapidly growing young knots have a soft ,pulpy texture and become covered with a velvety, olive-green growth of the fungus. In summer, the young knots turn darker and elongate. By fall, they become hard, brittle, rough and black.

During the following growing season, the knots enlarge and gradually encircle the twig or branch. The cylindrical or spindle-shaped knots may vary from one-half inch to a foot or more in length and up to 2 inches in diameter. Small knots may emerge from larger knots forming extensive galls. After the second year, the black knot fungus usually dies and the gall is invaded by secondary fungi that give old knots a white or pinkish color during the summer.

Smaller twigs usually die within a year after being infected. Larger branches may live for several years before being girdled and killed by the fungus. The entire tree may gradually weaken and die if the severity of the disease increases and effective control measures are not taken.

Most popular American plum varieties are highly susceptible to this disease. There are somewhat less susceptible varieties on the market including most Japanese varieties.

Remove all wild cherry and plum stands from any potential plum orchard location. Be aware of new wild growth in your orchard locale and remove them when they are spotted.

Established orchards or backyard trees should be scouted or examined each year for the presence of black knot, and infected twigs should be ruthlessly pruned out and destroyed by burning before bud break. It is very important to prune at least 2–4 inches below each knot because the fungus grows beyond the edge of the knot itself. If pruning is not possible because knots are present on major scaffold limbs or the trunk, they can be removed by cutting away the diseased tissue down to healthy wood and out at least 1/2 inch beyond the edge of the knot.

Fungicides can offer significant protection against black knot, but are unlikely to be effective if pruning and proper sanitation practices are ignored. In areas where the disease is highly present due on the homestead orchard or on a neighboring or abandoned orchard, protection may be needed from bud break until early summer. In disease prone areas, fungicides will provide the greatest benefit if applied before rainy periods, particularly when temperatures are greater than 55 degrees F. In evaluating black knot disease on your trees, remember that knots often do not become apparent until the year following infection.

Harvesting plums

Plums can be left to ripen on the tree

and harvested when they reach their mature size and color. If they are rich and sweet to the taste, and feel slightly soft when squeezed they are ready to be harvested. Plums can be stored for a few days in the refrigerator but will deteriorate in quality rapidly if held too long. Plums can also be frozen whole for later use in pies, jams, and other cooking.

Cherries

Whether your goal is sweet cherries to eat fresh in mid-summer, or tart cherries to freeze for a mid-winter cherry pie, you'll want to add a couple of these beautiful and productive trees to your homestead orchard.

Both sweet and sour cherries are stone-fruits that make wonderful jams, jellies, and preserves, as well as being useful in many baked goods and dessert recipes.

The sweet cherry produces trees up to 40 feet tall. Sweet cherries are somewhat more tolerant of low temperatures than peaches, and in favorable environments the trees may live for more than a century. Sweet cherry trees have fewer, but larger leaves than tart cherry trees. The fruits are large with a deep stem cavity, and can vary in color from yellow to red to purplish black. The stems or pedicels are about 1.5 inches long. The flesh ranges in texture from tender to firm, and is sweet and juicy. Most sweet cherry varieties are consumed as fresh fruit, but some varieties are used in making maraschino cherries.

The tart cherry, also called "pie cherry" or "sour cherry" produces a tree that is relatively small, up to 20 feet tall. The relatively thin limbs tend to droop and the tree has a bush-like appearance. Tart cherry trees are very cold hardy, but are shorter lived than are sweet cherry trees. The fruit of the tart cherry is relatively small, and has a higher acid and lower sugar content than sweet cherry. Tart cherries are used for processing, jam, and pie filling. Tart cherry is the latest blooming of the stone fruits, therefore would be less prone to frost damage than would sweet cherry.

The Duke cherry is a hybrid between sweet cherry and sour cherry. Most varieties more closely resemble sweet cherry although there are numerous variations in Duke cherries.

All commercial varieties of tart cherry are self-fertile, and there is no need to plant more than one variety.

Pollination requirements for sweet cherries are complicated by the fact that most commercial varieties are self-sterile, and some combinations are incompatible with one another, so it is critical to select varieties that are compatible. In addition, good pollinating varieties must bloom at the same time. Self-fertile varieties, such as 'Stella,' 'Lapins,' and 'Sweetheart,' are recommended for home orchards because they do not require cross-pollination and they can be used as universal donors to pollinate all varieties.

An example of the difficulty with sweet cherry pollination is shown here. Kristin is a self-sterile variety, and requires another variety for pollination, however, Kristin and Ulster are inter-sterile and should not be used to pollinate each other. Hartland, on the other hand, is self-sterile, but it is compatible to use as a pollinator with Ulster, Cavalier, and Kristin.

The easiest way to get around the confusion of sweet cherry pollination requirements is to plant a universal sweet cherry pollinator with your desired sweet cherry trees. Some of these universal pollinators are listed along with other varieties in the following section.

Tart Cherries

Self-fertile trees that do not require pollinators. You can select one variety and plant one or several trees of that chosen variety or you can plant tees of several different varieties.

- Montmorency cherry trees are productive and the fruit are relatively large, bright red, with white, firm flesh and clear juice. This is the standard tart cherry variety grown commercially in North America. Ripens in late June.
- Northstar cherries are medium sized, mahogany red fruit with red juice. Trees are small, which makes them easy to cover with bird netting. Ripens in late June or early July. The trees possess some resistance to leaf spot and brown rot.
- Meteor has medium sized, bright red fruit that ripens about a week after Montmorency. it is tart with clear juice, and semi-firm flesh. The variety is not acceptable for commercial processing because the pit is oddly shaped, and sometimes shatters during pitting. This is not a problem for home production and it can extend the harvest season because it ripens about 5 days after Northstar. Trees bloom later than Montmorency and Northstar and have some resistance to leaf spot.
- Surefire is late blooming, so is less susceptible to spring frost. The

fruit is bright red, medium in size, firm, and very tart.

- Danube ripens a little earlier than Montmorency. The fruit is dark red, medium to large in size, and sweeter than most tart cherries. This variety is widely grown in Europe.

Dark Sweet Cherries

- Lapins is a smaller tree that is renowned for the quality of its dark red, very juicy fruit that are rich in cherry flavor. Lapins has good split resistance, and is also self-fertile, making it a good universal pollinator for other sweet cherries. Lapins ripens early to mid-June.
- The Bing Cherry is considered one of the finest commercial sweet cherries available. This very large cherry ranges in color from a deep, dark red, to nearly black. The skin is very smooth and glossy, with a firm flesh and sweet flavor. Bing cherries are great for cooking as well as eating fresh. The flesh is firm, reddish purple in color and is extremely juicy and flavorful. Needs a pollinator.
- Cavalier is a black cherry variety that ripens very early, around the first week of June. The fruit is me-

dium to large, dark red, and has dry, firm flesh with good flavor. cavalier is resistant to cracking. Needs a pollinator.
- Black Tartarian is an old variety. The fruit are purplish black, small to medium in size, and heart-shaped. The flesh is dark red, sweet, soft and juicy. Not a good variety for crack resistance. Needs a pollinator.
- Hartland ripens around June mid-June. The tree is productive, but the medium sized, round, purple fruit is fairly susceptible to cracking. The fruit is of medium firmness, and has good flavor. Needs a pollinator.
- Kristin produces vigorous, high yielding trees. The fruit is large, aromatic, firm, and sweet. The dark red cherries are attractive, and moderately resistant to cracking. Fruit can be used for fresh consumption as well as for processing. A good tree for homestead planting. Needs a pollinator.
- Ulster produces high quality fruit that is large and sweet, with dark red flesh and a crisp texture. It is also moderately crack resistant, and ripens around mid-June. Ul-

ster is good for fresh consumption as well as processing. A good choice for the homestead orchard. Needs a pollinator.

- Stella produces large, black, heart-shaped cherries that are sweet, with medium firm flesh. Unfortunately it is also susceptible to cracking. It ripens around mid-June and is self-fertile. Stella makes a good universal pollinator for other dark, sweet cherries and is a good choice for the homestead orchard.
- Hedelfingen is an old European variety. It has black fruit that is medium to large, firm, and of high quality. It is also more crack resistant than most other varieties. The fruit is good for fresh consumption, freezing and processing. This variety ripens around mid to late-June. Hedelfingen is another cherry that is ideal for the homestead orchard. Needs a pollinator.
- Hudson is a late-season variety, ripening around the end of June. The fruit is dark red, medium to large in size, sweet and very firm. The trees bloom late, are vigorous, and are moderately productive, but may come into production late.

Fruit is very crack resistant. Needs a pollinator.

Light Sweet Cherries

- Whitegold is an early to mid-season, self-fertile cherry. The fruit can be used for fresh or processing purposes. Whitegold can serve as a universal pollinator for other sweet cherry cultivars. Fruits are yellow with a red blush. Reported to bloom later than other white-fleshed cherries and to have good field tolerance to bacterial canker and leaf spot. A good choice for the homestead orchard.
- Napoleon, also called Royal Anne, has pale yellow fruit with bright red cheeks. The fruit is medium in size, firm, sweet, and juicy, with fair fresh eating quality. Napoleon doesn't deliver consistently high yields, but it has fairly good resistance to cracking. Needs a pollinator.
- Rainier fruit is very large and sweet; the skin is yellow with quite a bit of red blush. The fruit is firm and the juice is clear. Rainier cherries are good for brining or fresh consumption. Studies show it has a low to moderate susceptibility to

cracking. Needs a pollinator.

Planting guidelines

Cherry trees generally do best where summers are not too long and hot or where winter temperatures are even and moderately cold. When planning your cherry tree planting site, remember that frosts and freezes during early spring, before bloom, may kill flower buds. Plant cherry trees on a site that is higher than the surrounding land, so the cold air can drain into the lower areas. Avoid planting cherry trees in any low spots or frost pockets on your homestead.

Cherry trees thrive in a wide range of soil types, so soil characteristics are much less important than cold air drainage. In general, the soil should be well drained, with a pH of about 6.5. Soil fertility and pH can be amended with fertilizer and lime applications.

Cherry roots are extremely sensitive to excessive moisture, which can stunt tree growth or may even kill the entire tree. Tree losses caused by the soil-borne fungus, such as crown rot or collar rot tend to be greater in wet or poorly drained soils. Be sure the soil in your desired site is well drained.

Dig planting holes just deep enough to allow the roots to be covered with 4 to 6 inches of soil, and about an inch wider than the original pot or root ball. Be sure to place the graft union about 3 inches above the soil line. Plant trees firmly and stake securely if needed. Water the new trees in well immediately after planting. Prune the previous season's growth on the leading branches by about half, and the side shoots to about 3 inches. Add a thick mulch layer to the soil surface over the root area to conserve moisture and reduce weed populations, but be sure to keep the mulch from touching the tree trunk.

Tart cherry trees are relatively small and can be planted 14 to 18 feet apart within the row, with 20 to 22 feet between rows.

Sweet cherry trees on vigorous rootstocks should be planted 16 to 20 feet apart within the row with 22 to 26 feet between rows. Sweet cherry trees on dwarfing rootstock can be planted 10 to 14 feet apart within the row with 16 to 20 feet between rows.

Maintenance

Water the tree every 14 days unless at least 1 inch of rain fell since the last watering. Fertilize with a complete fertilizer or fruit tree formula twice, once about 2 weeks after planting and then again 6 weeks after planting. To avoid root injury, place fertilizer in a band 6 to 8 inches from the trunk around the tree. Trees require high levels of nitrogen early in the season and lower levels are desirable during the late summer. Trees that are given access to high levels of nitrogen late in the summer will be less cold tolerant during the winter months.

Sweet cherries fruit chiefly on the spurs formed on the older wood. Sour cherries develop on shoots formed the previous season. Pruning of both sweet and sour cherries consists of maintaining the tree in an open habit with an evenly balanced head, together with the removal of any dead, diseased, crossing and rubbing branches. This minimal pruning is all that is necessary and should be confined to the spring only.

If birds are a problem around your homestead, you can use netting on the trees to protect the fruit from hungry birds.

Precipitation, heavy fog, or dew just before harvest can cause cherry fruit to crack. For some varieties in some years as much as 90% of the fruit may crack. Cracking is caused by absorption of water through the fruit skin. As water moves into the fruit, the fruit swells until it eventually bursts. Even a slight crack can serve as an entry point for the fungi that cause fruit rot. Fruit cracking is one of the primary factors limiting sweet cherry production in the eastern US. Cracking is most severe on young and lightly cropped trees. Varieties with soft fruit tend to be less susceptible to damage. Tart cherries are less susceptible to cracking than sweet cherries.

Insect pests and diseases

The Japanese beetle is a devastating pest of urban landscape plants.

Bacterial canker is a disease of cherry trees that is caused by a bacteria. This disease can kill or seriously damage your cherry trees. There are some cultural practices that can minimize infection. Other cultural practices may encourage infection and should be avoided. The bacteria is active in all phases of the growing season but is most devastating to young trees which are most susceptible to damage. The bacteria over-winter in bark tissues, and in apparently healthy buds or in the tree's vascular system. In the spring, especially when the conditions are cool and wet, bacteria multiply and emerge from their over-wintering sites and are usually disseminated throughout the orchard by wind and rain. Entry sites for bacteria include natural openings from flowers, etc, wounds, pruning cuts, or winter injury sites. The bacteria can infect flower buds and spurs. It can kill new spurs and leaves and then move into the trunk.

The cankers will sometimes increase in size for several years. If trunks or scaffold branches of young trees are infected, a canker may develop that may weaken or even kill the tree.

Cankers typically ooze amber-colored gum and often become entry sites for borers. Bacterial canker is very difficult to control. In the spring apply 1 to 2 applications of Bordeaux mixture before bud break. This mixture contains copper and sterilizes the tree surface. To minimize opportunity for this disease, do not dormant prune cherry trees because the wounds heal slowly. When working on or near trees that show evidence of the disease, sterilize pruning tools with Clorox after each cut. Apply an antibiotic, such as streptomycin, just before pruning in the spring and early summer. When mowing or cultivating around the tree, take care not to damage the bark because the wounds can become entry sites for the bacteria.

Several species of animals may feed on various parts of cherry trees or on the fruit. The severity of the problem varies around the state. Below is a brief discussion of the most common problems and possible control measures.

- White tail deer—Deer populations are increasing rapidly throughout the eastern United States. Deer coexist very well with humans and they often feed on garden plants including fruit trees. Deer seem particularly fond of sweet cherry trees. Deer browsing on young succulent shoots of sweet cherry trees can be so severe that the trees may never produce fruit or the trees may die. Bucks can partially debark a tree or break branches by rubbing their antlers against the trees in the fall. When the deer population is relatively low, taste repellents such as hot sauces and RopelTM work fairly well, but they must be applied every 7 to 10 days and after rain. Hanging bars of soap or small cloth bags of unwashed human hair or blood meal are effective odor repellents. In areas with high deer populations, repellents are not effective. In such areas, the only effective method is exclusion. Surrounding individual trees with wire fence or surrounding a planting with an electrical fence will help keep deer away from trees. These exclusion methods are often not completely effective, but they usually deter deer feeding enough so the trees grow and produce fruit.

- Rabbits—Rabbits may feed on bark just above the ground and, if damage is extensive, the trees may be weakened or killed. Surrounding the trunk with hardware cloth to a height of at least 12 inches above ground usually prevents rabbit damage.
- Voles—Voles are small rodents that feed on plants and are often confused with mice. Voles feed on roots of plants, and they can feed on roots or bark on cherry trees. Low populations can be controlled by placing household mousetraps in runs or on the ground around the tree. Make the ground under the tree undesirable for voles by controlling the vegetation under the tree and do not place organic mulches under the tree.
- Birds—Sweet cherry fruit is a favorite food of several bird species. Just as the fruit are ripening, birds can eliminate the entire crop in one day. Birds also feed on tart cherries, but they usually eat less than 70% of the crop. Various types of noisemakers are moderately effective for scaring birds. Placing scarecrows near trees, hanging aluminum pie plates that rattle in the

wind, placing large balloons with an eye painted on the side, or placing plastic snakes in the trees do little to scare birds. Birds eat fruit in large cherry plantings, but there are so many cherries that growers still harvest a sizeable crop. The only effective control for birds in small cherry plantings is with a physical barrier such as covering the tree with netting.

Cherry trees are susceptible to low temperature injury. The biggest temperature stress is warm sunny days followed by rapid cooling in the evening, which can injure the bark on the trunk and lower sections of branches.

During the winter the angle of the sun is low, and the temperature of bark on trunks and branches may be well above the air temperature because there is no foliage to provide shade.

On clear winter nights, especially when there is snow on the ground, air temperatures may drop very fast and cause bark injury, especially on the southwest side of the trunk, on the upper surfaces of relatively flat branches, or in narrow branch crotches. This type of injury is commonly referred to as "southwest injury" and sometimes causes bark splitting.

If bark splitting is discovered within a day or two of occurring, before the bark dries out, tack down the bark against the tree with roofing nails to repair the damage. Paint the damaged surfaces with an asphalt-based tree coating material to prevent drying and improve healing of the bark.

Often low temperature injury is not apparent for several months after the injury, or until the dead bark appears as a sunken area. By then there is nothing that can be done to improve healing. These areas of dead bark become entry sites for borers and disease infection.

Southwest injury can be minimized by painting trunks, especially on the southwest side, with white latex paint during the early fall. The white paint reflects the sun's heat away from the bark and minimizes the rapid temperature fluctuations that cause injury.

After leaf fall during the first, third, and fifth years after planting, paint the trunk from the ground to the lower branches, the lower branch crotches, and the lower sections of scaffold branches with a solution of half white latex paint and half water. Never use oil base paint because the oil will injure the bark. Paint can be applied with brushes or with a sprayer. A fairly efficient method of applying paint involves inserting a latex or rubber glove inside a cotton glove. Dip the gloved hand in the paint and rub your hand over the trunk and branches. Older trees are less susceptible to southwest injury, possibly because the bark is thick enough to resist injury.

Printed in the United States
149154LV00006B/18/P